Responses to Andrea Olsen's *Moving Between Worlds*

Eeva-Maria Mutka
Pen Pynfarch, Wales (2015)
Body and Earth: Seven Web-Based Somatic Excursions
Still image from videography
by Scotty Hardwig

"*Moving Between Worlds* is a timely addition to Andrea Olsen's distinguished sequence of books exploring the potential of embodied intelligence as a resource for personal, social, and planetary health. Throughout her career as a dancer and a teacher, she has connected art, physiology, environmental stewardship, and multi-cultural communication in ways that are at once utterly original and urgently necessary."
—JOHN ELDER, professor emeritus of English and Environmental Studies at Middlebury College and the Breadloaf School of English, author of *Picking Up the Flute*

"Andrea Olsen's writing calls us toward a critical quiet, an urgently quiet practice. Her guidance gives us the framework we need to regard our place—our bodies, our spirit—in this world. She has shaped a month of days to practice, remember, and remind us of what is within reach."
—BEBE MILLER, professor emeritus of Dance, Ohio State University, and artistic director Bebe Miller Company

"In our cautious, distanced modern milieu, we too easily forsake the body. Here Andrea Olsen reminds us that the body, our bodies, are the center of any engagement, interaction, or offering we will ever have. She brings multiple lenses to a 'scholarship of the body,' of lived experience, applying her findings to our needs for communicating across divides. This book is like no other, stretching between dance and dialogue."
—LAURA SEWALL, author of *Sight and Sensibility: The Ecopsychology of Perception*

"'What does the body have to do with communication?' one of Andrea Olsen's students asks impatiently. This book is her response, and it is a gift to us all. This fascinating, compelling collage of elements gradually reveals what makes the body so profoundly communicative, within oneself, with others and with the world. The body's complexity is reflected in the extraordinary richness of her means, encompassing science, stories, art, images, memories, philosophy, spirituality, and, above all, exercises that ground readers in their own felt experience. Andrea writes not just about the body but from the body. *Moving Between Worlds* is her body communicating with our bodies."
—ROSALYN DRISCOLL, author of *The Sensing Body in the Visual Arts: Making and Experiencing Sculpture*

"This book goes beyond simply making us more aware—it helps us to understand that communication, to be effective, needs to be open-ended, emergent, and ongoing. Olsen offers practices to prepare ourselves to communicate and to tune our actions in real time. In her expansive ways of thinking, she includes communication with the more-than-human world and all the beings that inhabit it. Reading this book is not only timely, it is urgently relevant as we meet both the challenges and moments of exquisite beauty in contemporary living on an endangered planet."
—CHRIS AIKEN, dance artist and associate professor of Dance, Smith College (from the Foreword)

"Working in the field of international relations requires an embodied and ethical awareness of cultures, norms, and values, as well as specific communication skills. Olsen quite sensibly puts the body at the front and center in her quest to understand the divides that can separate or connect us across perceived borders and boundaries. If you work or travel internationally, this book is a quintessential companion on the journey and can help you navigate your relationship with the world and your personal story, including self-awareness and self-managing stress levels."
—TANGUT DEGFAY, international development and public policy specialist, Addis Ababa, Ethiopia

"*Moving Between Worlds* will benefit many people involved in international communication! I can't wait to hold it in my hands. As we become more aware of our body and our senses, our perceptivity deepens. We participate in the world of creativity with music, with dance, and with poetry, connecting globally beyond perceived linguistic and cultural barriers."
—NÜKHET KARDAM, author of *From Ottoman to Turk and Beyond: Shimmering Threads of Identity*, and co-creator (with Amir Etemadzadeh) of the Rhyme & Rhythm Project

"The movements that our bodies make are integral to our spiritual lives. Patterns of bodily movement train our senses; we grow willing and able to perceive greater realities of which we are an active part. By providing shining examples from myriad traditions as well as thoughtful guidance in specific exercises, Olsen offers us ample resources for transforming our relationships to ourselves, to others, and to the Earth, so that we may participate more consciously in the worlds our movements are creating."
—KIMERER LAMOTHE, philosopher, dancer, scholar of religions, and author of *Why We Dance: A Philosophy of Bodily Becoming*

"Feeling at home in our bones and bodies allows inherent intelligence to keep flowing throughout our embodied selves. As we inhabit our full sensory potential with internal integrity and extend awareness to all that surrounds us, we allow rhythms inside to meet rhythms outside. Olsen's book offers multiple embodiment practices that support this process, clarifying the dynamic exchange that is always happening in the day-to-day experiences of living and communicating."
—CARYN MCHOSE AND KEVIN FRANK, founders of Resources in Movement and authors of *How Life Moves: Explorations in Movement and Meaning*

"Embodiment skills allow us as performers and as humans to tune into the frequencies of others, no matter what city or country we're in. The invaluable tools for inherent nonverbal ways of communicating that Olsen shares through her stories, research, and movement practices are perfect for all those who are looking to connect more deeply—and for those who feel we are all already connected."
—CAMERON MCKINNEY, artistic director of Kizuna Dance, and 2019–20 U.S.–Japan Friendship Commission Creative Artist Fellow

"Olsen's dynamic, interactive guide offers essential body-based education and inspiration for us all—including teens and college-level adults. Her somatic activities can help them to cultivate body awareness and develop the embodied resilience needed to support healthy communication, whether within their family, during international travel, or in their lifelong journey as moving bodies on the Earth. If you'd like to support your teen or college-aged adult to become more self-aware, resourceful, and communicative—read and share this book!"
—SUSAN BAUER, founder of Embodiment in Education and author of *The Embodied Teen*

"Ooh wow yummy book! Reading felt like a mug of warm tea nestled between hands. Some very helpful reminders here for the process of studying abroad. Arriving somewhere—to this place, to this time, to this growing body—requires real shapeshifting. I can already feel this wanting to be re-read and breathed into practice. So, thank you!"
—HANNAH LAGA ABRAM, Dance/Anthropology student at Middlebury College currently studying in Ireland

MOVING BETWEEN WORLDS

A Guide to

Embodied Living and

Communicating

ANDREA OLSEN

WESLEYAN UNIVERSITY PRESS

Middletown, Connecticut

Wesleyan University Press

Middletown CT 06459

www.wesleyan.edu/wespress

Unless otherwise noted, text and compilation © 2022 Andrea Olsen

Foreword © 2022 Chris Aiken

Body-Mind Centering® is a registered service mark and

BMC℠ is a service mark of Bonnie Bainbridge Cohen

Manufactured in the United States of America

Designed by Richard Hendel

Composed by Rebecca Evans in Utopia and TheSans

Library of Congress Cataloging-in-Publication Data

available at https://catalog.loc.gov/

paper   ISBN 978-0-8195-8089-4

e-book   ISBN 978-0-8195-8090-0

5 4 3 2 1

This book is dedicated to creative colleagues
PETER SCHMITZ, NANCY STARK SMITH, *and*
GORDON THORNE, *honoring the depth and*
continuity of connection, unique and precious,
through four decades of making

Poppy Study #1 by Kristina Madsen
Maple, milk paint, gesso
Photograph by Stephen Petegorsky

Contents

Helmet Mask with Praying Mantis
(before 1914)
Papua New Guinea, New Britain, Sulka
Wood, paint, 43 1/2 in.
Gift of Evelyn A. J. Hall, 1981 (1981.331.1),
The Metropolitan Museum of Art, New York,
NY, USA © The Metropolitan
Museum of Art; image source: Art Resource, NY

IV ENHANCING AUTHENTICITY

Foreword

CHRIS AIKEN

In *Moving Between Worlds: A Guide to Embodied Living and Communicating*, Andrea Olsen writes with great wisdom about the relationship between embodiment and communication, drawing on her many years of dancing, performing, writing, teaching, and collaborating with artists, scholars, scientists, therapists, healers, and community activists. In her previous works *BodyStories*, *Body and Earth*, and *The Place of Dance*, Olsen has been a leader in advocating for meaning and knowledge-building through the development of embodied intelligence. For her, this has always included a weaving-together of awareness with the biological foundations of our being. She braids these subjects together in ways that are quite accessible: the ecological, the poetic, the personal, and the cultural strands of our lives.

In this book, Olsen states: "Recognizing the impact of perception may be the most essential component of communication." Having studied and explored perception for many decades, I could not agree more. In the field of ecological psychology, founded by J. J. and Eleanor Gibson, action is viewed as inseparable from perception. Without the ongoing ability to monitor the progress of our actions, affordances of our environment, and the emergent actions of human and non-human beings, we would not be able to survive, let alone communicate with one another. To communicate effectively involves listening to the self, as well as others, to tune one's attention. Listening offers ongoing feedback that can be used to change one's perceptions and tune one's actions.

Each chapter or "Day" is full of practices and pathways that readers can explore to develop sensitivity and relational agency within their perceptions. Olsen encourages the reader to slow down, or even to pause what they are doing, to allow new information to surface and to recognize patterns in the ways we perceive, patterns that not only shape our body and our movement but also our beliefs about and connections with others. Her work here is very much connected to the pre-movement anticipatory states described in the research of Hubert Godard, Bonnie Bainbridge Cohen, and Nikolai Bernstein. Unconsciously, we are continually readying ourselves to act by orienting our senses and toning our tissues to what we anticipate is about to happen or what we are about to do. As Olsen points out, our preparations are not always connected to what is actually happening but are the result of our education, culture, and personal history. She reminds us that the speed of perception and action is so fast that it is very difficult to refine and adjust our actions once they have begun.

Through detailed exercises and stories, Olsen offers us tools to notice and explore how our perceptions of self, others, and the world are interconnected and foundational to our capacities to communicate with one another. This book goes beyond simply making us more aware: it helps us to understand that communication, to be effective, needs to be open-ended,

Chris Aiken
Photograph by Chris Randle

emergent, and ongoing. Olsen offers practices to prepare ourselves to com-municate and to tune our actions in real time. In her expansive ways of thinking, she includes communication with the more-than-human world and all the beings that inhabit it. Reading this book is not only timely, it is urgent. As Olsen reminds us, we need our bodies to be fully present to meet both the challenges and moments of exquisite beauty of contemporary living on an endangered planet.

Entryways

An Introduction

*A bow of respect toward the material,
a bow of respect toward the reader.*
—*Barry Lopez, from "The Wild Road
to the Far North"*

Andrea Olsen
London, England (2015)
*Body and Earth: Seven
Web-Based Somatic Excursions*
Still image from videography
by Scotty Hardwig

Moving Between Worlds is the last of my quartet of books on the body, combining science and scholarship with four decades of direct experience as a dance artist and educator. My work draws on the multidisciplinary body of research and writings that feeds what is now called the field of *somatics*, and for this reason, there are many different ways to enter the text. The foundation of this work is in dance, human anatomy, and Authentic Movement. My work also draws on ecological psychology, social anthropology, and neuroscience as well as linguistics and the literature of mindfulness.

The book you are reading is focused on the experience of communication. Like my prior offerings, it describes a series of practices that bring together evolutionary movement and experiential anatomy to clarify and orient the body; perceptual explorations to understand the nervous system and restore resilience in alignment; and the disciplines of writing and specific movement practices to integrate and articulate our findings.

ORIGINS

This book is based on a life lived dancing. Every word is filtered through the body, my body. If I haven't felt my way through a topic, I don't write about it—like breath and voice, or rest and restoration. Speaking from the body, not just about the body, is part of somatic awareness. This perspective honors the deep intelligence of the body as subject, not object; the Earth as subject with intrinsic intelligence and integrity; and both as interconnected with much to teach us if only we learn to listen.

BETWEENNESS is a central theme. We wake in the morning, move from silence to speaking, indoors to outdoors, private to public, and personal to professional with multiple transitions. As we communicate with different parts of ourselves, fellow humans, and our surroundings, we cross thresholds from one dimensional realm to another. Amidst all these transitions, not-knowing is constant: there is a gap between where we were and where we are going. Interpretation of each present moment is based on our past experience and on the health and receptivity of our bodies. In this state of betweenness, we are immersed in fresh possibility. As we inhabit this "ecotone," or terrain of potential creativity, noticing what brings personal joy and insight is part of the unfolding process.

Images are what make my heart sing. Juxtaposed with themes, they support intuitive connections. A visual artist by training and family heritage, I experience art images as home ground in the realm of nonverbal communication; visual art offers a way of knowing that enriches and enlivens beyond conscious thought. As Barry Lopez writes in his vivid autobiographical narrative *Horizon*: "Art's underlying strength is that it does not intend to be literal. It presents a metaphor and leaves the viewer or listener to interpret." When encountering a compelling visual, I feel unmediated joy. And visual artists, in my experience, are essential colleagues: they see.

Embodied communication is a learning process. When I am interviewed for newspapers or podcasts, I stumble to find words that match my experience. What feels coherent and easeful inside my thinking mind is revealed as awkward. I write to learn: if I were already a great communicator, why bother to spend decades of life investigating and practicing? I danced because it was silent. I wrote because I could edit my words. I spoke on stage while dancing because I could rehearse each phrase. I now film and photograph to compare what something "feels like" to what it "looks like." And when I write a book, I simultaneously make an hour-long dance to check if what I am saying feels true.

This writing project grew out of experiences bringing embodied perspectives into graduate courses on global communication at the lively Middlebury Institute of International Studies at Monterey, in California. Career-focused students from many countries sat behind desks (on rolling wheels) with computers and cell phone screens between us. Generally, they expected fast, relevant, easily applicable tools, accompanied by PowerPoint presentations. How is embodiment communicated in this context? How can cutting-edge research on perception and brain-body interconnectedness be relevant? How can practices for self-managing

Embodiment

When asked, "What does it mean, 'to inhabit the body'?" dance artist Nancy Stark Smith responds: "For me, inhabiting my body is about feeling the actual sensations of my exact body in this exact moment—for example, expansion and release of the chest with breathing, crick in the left trapezius muscle. It is about being present throughout your body as it is, not just having one to carry you around."

Interconnectivity

We were all just "bumbling about," Janet Adler says about our time together in the 1970s to 1990s in Northampton, Massachusetts, "focused on our work." Janet and the Discipline of Authentic Movement are influences that underlie this book. Collaboration and interconnectivity are not all about talking, discussing, deciding, and planning. In fact, they have much more to do with the "nonverbal thing"—the betweenness of people, ideas, and images experienced and recorded through the living, moving body. We are impacted at every second by context and companions, even if we don't notice at the time.

Spinning

Growing up on a farm in the midlands of the U.S., I would stand in our field spinning and see the horizon swirling 360 degrees around me. From there I could go anywhere, any direction, and that became my aspiration: to have a home tether but go far afield, meet other people, other cultures, other views. Dance was my partner in this goal. Ballet classes in the nearby city made spinning into art, and art could take me anywhere—and did.

stress and working with trauma in disaster-relief zones be communicated not as concepts but as embodied skills? A book that included science-based embodiment principles and practices for living and communicating became necessary.

Embodiment skills have become increasingly relevant in the ensuing years amid the Covid-19 pandemic, global racial protests, and political challenges. The word embodiment is "here, there, and everywhere" in books, articles, songs, online summits, and social media conversations that cross time zones, continents, and academic disciplines. My desire in this book offering is for everyone to have embodied knowledge and a daily movement practice to feel connected to place, to people, and to ancestors—human and other. There is much to attend to, and our choices are impacted largely by our states of mind and attention. Perhaps a guidebook can be of use as we navigate the increasingly complex and noisy terrain of our lives. We are in uncharted waters, moving fast. How do we become calm, find flow, feel space and spaciousness, and appreciate the unique and universal bodies we inhabit?

LOOKING BACK

For those interested in the scholarly ground that underpins this book, we can reflect on several authors, scholars, and creative investigators. Current understandings of embodied knowledge from within psychology and social anthropology can be traced back to the tradition of American psychologist J. J. Gibson (*The Ecological Approach to Visual Perception*, 1979), who posits that the world is full of rich, higher-level information. Becoming an expert means refining (refreshing) our senses to perceive what is meaningful and attuning to finer and finer differentiations of meaning. Social anthropologist Tim Ingold (*The Perception of the Environment*, 2000) further developed this notion, recognizing the dynamics of perception and perceivability. He states: "Skills are grown, incorporated into the human organism through practice and training in an environment." Active engagement attends to "what it means to 'dwell,' and on the nature of skill," weaving together approaches from social anthropology, ecopsychology, and developmental biology.

Visual psychologist Laura Sewall (*Sight and Sensibility*, 1999) focuses on place-based seeing. She asks: "Why is it important that we see from an embodied perspective when communicating? We are always held within a context, and with our feet on the ground, somewhere. But if we don't place ourselves, if we don't know where we are standing, we will not be balanced in our interactions." Plant ecologist and indigenous author Robin Wall Kimmerer (*Braiding Sweetgrass*, 2015) concurs. How we perceive the world shapes our identity. We carry the landscapes and peoplescapes of our childhood homeland within us. If we leave our homeland, that early heritage accompanies us. She also notes that discipline-specific and science-specific terminology can create a "wall of words," separating us from rather than connecting us to the world we inhabit.

These writers are phenomenologists at heart: the body does not exist separately from the body-as-lived experience; a person does not exist separately from the environment but is embedded in it. Rolfer, researcher, and dancer Hubert Godard amplifies this view when he writes about the importance of the gravity response for efficient alignment, which affects the space around and between bodies. He describes the theory of tonic function and the importance of fascia in whole-body awareness and communication, emphasizing the pre-movement and its impact on tonal readiness. Godard's work informs Caryn McHose and Kevin Frank's explorations in phylogenetic development and resonance (*How Life Moves*, 2006). With effective perceptual orientation, reception increases. In other words, such preparations can help us in our ability to note what's going on and to move beyond our preconceived notions by allowing us to remain skillfully receptive and attentive.

Neuroplasticity and the capacity to change behavior through awareness has been articulated by Norman Doidge (*The Brain that Changes Itself*, 2007) and Sandra and Matthew Blakeslee (*The Body Has a Mind of Its Own*, 2008). Much of what has been intuited in the past is now documented scientifically, including the importance of the brain in the gut discussed by Michael Gerson (*The Second Brain*, 1999) and the microbiome discussed by Emeran Mayer (*The Mind-Gut Connection*, 2018). The multiple roles of the autonomic nervous system in the treatment of trauma have been explored through the development of somatic experiencing by Peter Levine (*Healing Trauma*, 1999); co-regulation as an interactive process engaging the social nervous systems of both therapist and client by Stephen Porges (*The Polyvagal Theory*, 2011); and the integration of advances in brain science, attachment research, and body awareness into treatments around traumatic stress by Bessel van der Kolk (*The Body Keeps the Score*, 2015). Gabor Maté (*When the Body Says No*, 2008) and Dan Siegel (*Mindsight*, 2009) share insights on addiction recovery and mind-body wellness.

Interculturalists Houman Sadri and Madelyn Flammia (*Intercultural Communication*, 2011) offer a comprehensive walk-through of theory and practice related to communication strategies in global contexts. Turkish-American political scientist Nükhet Kardam (*From Ottoman to Turk and Beyond*, 2016) delineates a more intimate familial view of intercultural and multi-ethnic identities linked to women's human rights. Resmaa Menakem (*My Grandmother's Hands*, 2017) addresses embodied racialized trauma and promotes "somatic abolitionism" as "living, anti-racist practice and culture building." Nature writer and cultural anthropologist Barry Lopez (*Horizon*, 2019) takes us on journeys to multiple remote sites, detailing the "long learning curve" to unlearn cultural assumptions and noting the urgency of global climate action.

Philosopher of religion Kimerer LaMothe (*Why We Dance*, 2015) introduces the idea of "bodily becoming" to illuminate the vital role that movement plays in the process of developing capacities that distinguish us in degree as human—including our abilities to empathize, cooperate, engage

Com Des

Showing visual artist and author Josie Iselin a draft of this book, she reminds me that Communication Design (Com Des) is a field of study. "How can you write a book on communication and not think about design?" she queries. "So much is said by the book layout itself." I nod my head in agreement. Form is content. In my books, the space on each page, the feel of the paper, the ideas juxtaposed with images create the context for the words.

California

When I met Nükhet Kardam over coffee at Café Lumiere on a sunny March day in Monterey, California, I knew I was not in New England. She was enthusiastic, effusive, generous with her ideas and gestures—exclaiming, "that's fantastic!" Her eyes were attentive, her hands defining the air. Her comments began as questions; her sentences consistently ended with "What do you think?" This meeting changed the trajectory of my creative investigations. It was an introduction not just to a person and a personality but to the power of cultural heritage, academic lineage, and West Coast fun.

in ritual, and adapt to multiple environments. Hillary Rodrigues and John Harding (*Introduction to the Study of Religion*, 2009) elucidate historical and critical contexts around spirituality and religious perspectives. Resources in mindfulness training through introspection and meditation in difficult times are articulated by Jon Kabat-Zinn and Thich Nhat Hanh (*Full Catastrophe Living*, 2013) and Pema Chödrön (*When Things Fall Apart*, 2000). Anthropologist and dance scholar Yvonne Daniel (*Caribbean and Atlantic Diaspora Dance*, 2011) offers several cultural lenses through which expressive dances release and channel physical and emotional stresses "to ideally create individual centeredness, communal harmony, and fierce, but loving protection throughout the local environment."

Practice *as* scholarship is a concept deployed by linguistic theorists as early as J. Samuel Bois and Gary David (*The Art of Awareness*, 1966) and has evolved into a flourishing subfield within art and performance theory. One example is the notion of PaR/Practice-as-Research, discussed by British scholar Anna Pakes (*Art as Action or Art as Object*, 2004) and Ben Spatz (*What A Body Can Do*, 2015). If we understand the levels of abstraction in language formation from initial stimulus to sensations, interpretation, meaning-making, conclusions, and theories, then we see that theory must constantly circulate back to "lower" sensory data, creating a cyclical balance between lived experience and scholarship. The "To Do" investigations in *Moving Between Worlds* are at heart a call to sustained research in embodied techniques.

Finally, Mabel Todd's pioneering work (*The Thinking Body*, 1937) created an important baseline for experiential ways of knowing and considering the body as subject, not object: not just any pelvis, but one's own pelvis—felt and inhabited in all its universality, uniqueness, and detail. This embodied perspective was carried forth in the writings of Bonnie Bainbridge Cohen (*Sensing, Feeling, and Action*, 1993) and Janet Adler (*Arching Backward*, 1995), whose insights are featured in this book.

MOVING FORWARD

Embodiment—consciously inhabiting the lived body—is a theme of our time. There are a number of programs and courses where this book can be especially useful, including:

A. Undergraduate or graduate-level survey courses on somatics, experiential anatomy, somatic psychotherapy, expressive arts, dance composition, and performance studies.

B. Environmental and international studies programs inclusive of whole-body and place-based awareness.

C. Communications courses that highlight the 90 percent of communication that is "non-verbal."

D. Psychology courses that focus on strategies for wellness, trauma recovery, and mindfulness.

E. Professionals in health centers; art, dance, movement, and physical therapists; and early interventionists, including youth educators and social and emotional learning specialists.

Note on repetition within the text

Teachers and students may find that some chapters are more applicable to their particular context. To accommodate single-chapter users, there is some intentional repetition of basic concepts and terminology, for example: the meaning of *body* (as a term inclusive of one's mental, physical, emotional, intuitive, energetic, and spiritual selves—however you define the parts); and *interconnectedness* (body systems and earth systems as one); along with my definitions of *communication* (the exchange of information, emphasizing the "gap of betweenness" that occurs), and *embodiment* (consciously inhabiting the lived body).

I believe conscious embodiment is central to our easeful living and communicating and that it enhances effective and joyful interaction. Change is possible and essential as we build on investigations of the past and move forward together in this challenging contemporary time.

Grand Lake Stream, Maine

2022

My Way
Bonnie Bainbridge Cohen says about my work: "You point toward and assemble." I think she is right.

Paul Matteson
Blue Barn,
Bramble Hill Farm,
Massachusetts
(2015)
Body and Earth:
Seven Web-Based
Somatic Excursions
Still image from
videography
by Scotty Hardwig

Preparations
Using the Text

Moving Between Worlds offers daily explorations in embodied living and communicating. Each of the thirty-one chapters or "Days" is a walk, stroll, or hike through a specific terrain. They can be experienced sequentially or in any order, over a month or several years, individually or within a group. Themes are addressed through an art image, factual information, personal anecdotes, and somatic excursions that invite the reader/viewer/mover to engage different learning styles. *Connections* paragraphs conclude each chapter, offering a daily invitation to bring the material directly into one's life. In several chapters, colleagues offer additional expertise and perspectives, incorporating conversations we have had during the writing of this book and over many years. At the end of thirty-one Days, each reader will have a kaleidoscope of experiences to deepen embodiment—clearing stress, building energy, and flowing with more ease and discernment through the ever-changing dynamics of daily living and communicating.

The book's themes are drawn from one person's journey, filtered through a lifetime of learnings and experiences. Rather than a comprehensive text on a single topic, this book includes perspectives and practices developed with students and fellow educators through decades of teaching and performing globally. In this way, each chapter addresses topics that have been lived in sufficient depth to share them with you, the reader.

Although embodiment is a birthright, some aspects of awareness can take time to unravel as you address multiple perspectives: personal, historical, cultural. Whether it is your first experience of an embodied practice or an exploration you have returned to again and again, the invitation is to notice what this material offers to you and how it can support each unique moment of your life. As you enter the realm of embodied awareness, you can stay present to "what now, what next, what if . . ."

LOCATION
Choose a private-enough place to investigate freely and without interruption, with a clean floor and enough space to move while standing or lying down. We spend so much of our days sitting or outwardly distracted by multitasking. This is an opportunity to focus, remove clutter, and connect inward.

"TO DO" EXPLORATIONS
Read through the prompts before you do them. You can also record yourself reading them aloud and play back your voice as a guide. The explorations invite repetition; they are practices, not one-time events. As you become familiar with the process, your body remembers the pathways and awareness deepens.

We all have our experience of the inner journey.—Janet Adler, from "Where Are We Now?" at the Somatics Festival 2019

Markings

Tangut Degfay from Ethiopia joins me in the process of investigating embodied living and communicating by hosting a spring series of events at the Middlebury Institute of International Studies at Monterey, in California. Describing the tribal marking on her forehead, she acknowledges that in moving from her rural community to the city, the marking was a detriment. But when she traveled to the United World College in Norway for high school, and then to college and graduate school in the U.S., the marking was seen as individual and unique. Her own attitude is what matters: Markings are about home, a reminder that she comes from some "place" and some "people," wherever she travels in the world.

Tangut Degfay
Monterey, California (2018)
Photograph by Stephen Keith

Urgency

Book project editor and former student Mandy Kimm emails from London, January 1, 2021: "So many messages from our cultural landscape encourage a kind of striving unconsciousness—keep going, keep buying, keep your head down. The PRACTICE of embodied awareness is about bringing the head up—spine long, belly soft, weight grounded, senses available—to navigate skillfully as our sweet, vulnerable, complex inner world is *in touch* with the vast outer world. Especially to someone at my age, in mid-twenties, there is an urgency to this. Every decision feels like *the* decision that will determine the course of a life, my life. How can we consciously choose to participate in forming the world we inhabit, to do the difficult and easily thwarted work of being vitally awake to the choice we have of how to live?"

JOURNAL NOTES

Identify a consistent method to respond to the writing prompts: in a note-book, on a computer screen, or in a drawing pad. This regular practice involves catching ideas and inspirations as they arise as well as notating your questions and curiosities. Even if writing feels awkward at first, keep going; journaling helps track your experience.

TIME AND TIMELESSNESS

Reserve a specific time in your schedule, and value self-nourishment. Mark your calendar, show up, and then explore at your own pace. If you have a tight schedule, set a timer so you don't have to think about clock time. There is a clear beginning and ending, but between those two markers is body time, where you meet unknown places in yourself.

VIDEO RESOURCES 👁

Videos are available for several chapters to support your investigations. Multimedia resources can add dimensionality to written materials. Watch for the "eye" icon that indicates web-based companion materials here: www.weslpress.org/readers-companions/.

UNDERLYING PATTERNS AND PERCEPTION

Aurora Borealis and Teepee
Blachford Lake, Northwest Territories, Canada[1]
Photograph © Kevin Schafer

Basic Concepts

Communication is inherent—every person communicates. Living and communicating with more ease and integrity through conscious embodiment is the subject of this book. Three basic concepts frame how we begin our journey:

Body is whole. In this book, the word *body* is an inclusive term involving all aspects of self. This includes our mental, physical, emotional, intuitive, energetic, and spiritual selves—however you define the parts. And everybody is included, all sizes, abilities, ethnicities, and ages. Rather than mind over body, we recognize whole-body intelligence. Embodiment is a process of inhabiting this full terrain with awareness—coming home to our bodies.

Body is part of Earth's body. In this book, the word *place* locates us in a specific landscape and peoplescape moment by moment. This includes the air, water, minerals, plants, and animals that are around and inside us. Rather than experiencing ourselves as separate, we recognize interdependence and interconnectivity. Embodiment of place is a process of extending our awareness to include the energies that surround us—fully engaging the places we inhabit.

Bodies communicate. In this book, the word *communication* means exchange of information; betweenness is inherent as signals pass from one source to another. Communication occurs within the body—neuron to neuron, hormone to target gland, system to system; and outside the body with other humans and the natural world. If you are living in a body, you are communicating—all the time.

Throughout, we recognize that living and communicating are interwoven like fine fabric. Awareness of authenticity, interconnectivity, and reciprocity is determined by the underlying tone of the body—our degree of resonance within ourselves and with the other people and places that inform our lives.

Connections

As you begin your exploration today, the invitation is to focus on the body—your body—to establish firm footing for embodied living and communicating. At various times during your day, pause and notice your whole self. Breathe in your connection to the world around and inside you; breathe out, and release restrictive tension. Enjoying this opportunity to muse and amuse without over-judging, you can notice how the three basic concepts within this chapter are relevant to living an energized and meaningful life.

Who is there big enough to love the whole planet?—E. B. White, from One Man's Meat

Why Body?

"What does the body have to do with communication?" This impatient question from a graduate student gives me pause. "I'm in a program for international development; if I wanted to study the body, I'd have gone to school for biology." I am teaching about breath and voice and think, "What about communication doesn't include the body?" Reading our escalating body tension, however, it seems best to affirm the question. There is pressure in graduate school to learn "usable skills" and stay focused on getting a job to pay the bills. Resistance is often a doorway; you just don't know what's on the other side. That question inspired this book.

Body Language

When looking for a book title that included embodiment, I checked internet sources for copyright conflicts. "Whole-body intelligence" and "embodied intelligence" were mostly linked to books on artificial intelligence and robotics. The understanding that intelligence requires a body is both visceral and visual. Comic strip illustrators utilize gesture and facial expressions in detail. So do gamers—choose your favorite avatar, and you inhabit a new way of being.

Self-Knowledge

Embodiment practices are generally lumped under "soft skills" in courses on communication. When I tell this to a friend who is a psychotherapist, she laughs and says decisively, "There is nothing 'soft' about self-knowledge. It takes so much work!"

TO DO

BASIC STANDING POSTURE

3 minutes

Let's begin by establishing a basic standing posture. Relaxation and grounding allow connection—to yourself, to other people, and to place.

Stand with your feet shoulder width apart, knees slightly bent. Feet are parallel (toes forward, heels back):

- Relax your arches and toes to reduce tension so you can feel the ground.
- Extend your tailbone down toward the earth.
- Release your arms to your sides and open the palms.
- Reach the crown of your head up toward the sky and free the back of your neck.
- Take three deep, full breaths, feeling the belly and low back expand with each inhale and condense with each exhale.

TO WRITE

THE STORY OF YOUR NAME

10 minutes

We are asked to say our names many times in introductions. Write the story of your name. And if you don't have a story around your name, write about that. Then read your writing aloud, hearing what you have said. Often, hearing and feeling your own words is different from writing. (5 minutes to write, 5 minutes to read and listen.)

Attitudes

Attitudes affect our actions and our interactions. What we think shapes what we do and how we construct our view of the world. Attitudes are based on our life experiences. They also reflect the state of our attention at any moment and the level of our confidence and ease in specific situations. If we are anxious and afraid, or secure and relaxed, interactions are impacted. Our attitudes about communication and which words and modalities we choose reflect our ways of engaging with life.

How are attitudes formed? We all have a worldview, whether we think we do or not. This is shaped by cultural and familial norms as well as personal values and environmental context. Perception plays a role, and interpretation of perception is based on personal experience. For survival, we need a baseline for responding, especially in fast-paced environments and pressured situations. This is called apperception—the mental process by which a person makes sense of an idea by associating it to a matrix of ideas already embedded. It is the process of understanding something. For example, we identify a chair as a chair because it is related to the chairs we have seen and touched before. This allows us to "make sense of" complex environments quickly and respond without excessive energy expenditure.

How do we interact with attitudes that limit communication with the self and others? Gender and race, which are endowed, and class, which is inherited, are familiar lenses that others use to identify us and that we may use to identify ourselves. But each of us has many other dimensions than our general external identity markers. Yes, I'm a white, middle-class woman, and I'm also a wife, stepmother, dancer, writer, daughter, sister; I'm moody, funny, and I can change.

EMBODIMENT

Attitudes about the body are central to communication styles and skills. In American culture, the body is often considered too superficial for serious study—reserved for athletes and dancers. Or it is viewed as too complex and delegated to medical professionals and scholars. Media exploitations often co-opt body awareness through a focus on sexuality and competition. Participants in movement practices generally feel the positive effects of movement on mental clarity and mood but are socialized to consider the body as less important than the mind. Movement is seen as an extracurricular add-on to serious, focused study, and self-care is the last thing on our daily "to do" lists.

Yet, the body is our primary means for communication and connection. The evolving discipline of neuroscience now confirms the degree to which body and mind map an interconnected whole.[1] Movement is our

We are the accumulation of our experience.
— Bessel van der Kolk, from The Body Keeps the Score

5

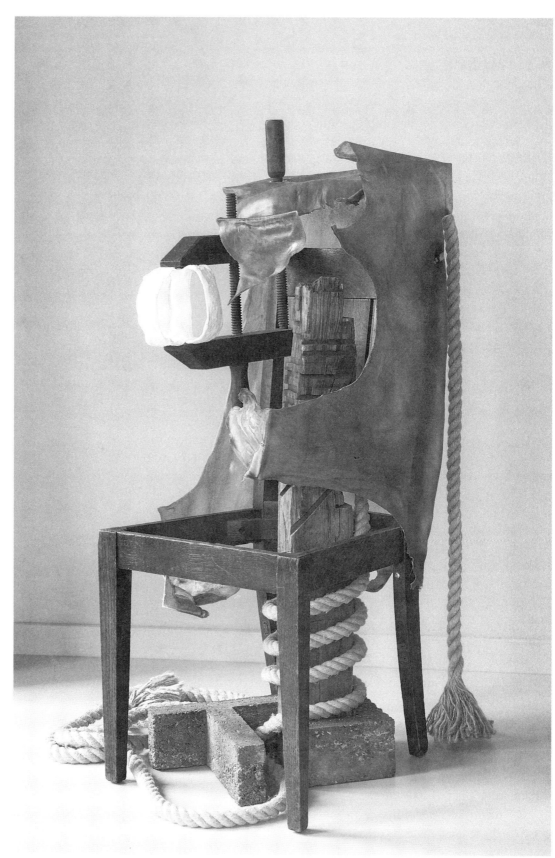

Conversation 3 (2010)
by Rosalyn Driscoll
Wood, rawhide, rope,
concrete, plaster,
45 in. × 25 in. × 30 in.
Photograph by
David Stansbury

first language. It underlies our capacity to communicate, connect, and feel emotionally what we are doing. We know that movement affects the mind, and the mind affects movement. Try to skip down the street and stay in a bad mood; it is almost impossible. The common language of the body is something humans share globally.

Sadly, many people in industrialized countries have lost or diminished *somatic literacy*—the skill of reading body language—because we have shut down perceiving bodily sensation as a way of knowing ourselves and the world. Visual and auditory learning are prioritized in many schools, and we lose the importance of touch, movement, and emotional connectivity in discerning relationships and meaning. We disregard, overlook, and leave behind our deepest ways of knowing.

Attitudes about the body extend to the larger body of Earth. Nature and science impact embodied communication skills. Noticing the underlying commonality of life on this planet is a good place to start as we meet others in intercultural and interdisciplinary contexts—we stand on the same ground, breathe the same air, drink the same water, and map both inner and outer dimensions of space and place. Colleagues and friends in different disciplines are increasingly essential to address the dimensionality of global challenges, sharing concerns and opening possibilities.

The arts as modes of communication are often seen as "extra"—something you do in your spare time or that are added in at the last minute to give liveliness to a presentation. Yet movement, visual images, videos, stories, music, and the fine art of culinary delights appeal to our heart's desire for nourishment at the most fundamental and sophisticated levels. Because they are multi-sensorial and immersive, they focus attention, motivate people to show up at an event or view an online posting, and help sustain concentration. Communication becomes dull or one-sided without shared space for exploring the "edge" around the attitudes that govern our lives through creativity and discovery.

Conflicting worldviews can create dead-end communication, or they can be avenues for relevant investigation and curiosity. Fixed attitudes often result in binaries—shortcut labels for describing and simplifying complexities: East vs. West, mind vs. body, male vs. female. Addressing reductive thinking and the overuse of acronyms and "isms" requires the capacity to stay oriented and be present to multiple views. As we look for commonalities and respect differences, we enhance our capacity for complementarity—the ability to perceive more than one perspective at a time.

Connections

This is a good day to orient your attention toward attitudes that shape your current life. Thought patterns affect who you are and who you might become. Remembering that the mind gravitates toward negativity, you can include within your day perspectives that support your deepest self. Try on a good, positive attitude and notice a few best moments throughout your day. Make a gratitude list at night before you go to sleep and wake up with

Perspective

Navigating back roads on two-lane highways in the American South in the 1950s was my childhood induction into cultural difference. What I took for granted to be true in Illinois no longer applied: that neighbors are your friends; that food is a common language; that children are happy. It was the era when bathrooms were still marked "colored" and "white"; chain gangs with men (usually black) were doing road work, wearing striped prison uniforms, connected by shackles and chains, overseen by men (always white) with rifles and accompanied by dogs. Children sat with bare feet dangling off rickety porches watching our car pass. We kids didn't know whether to wave or look away: there were no smiles. This annual journey stayed tucked away inside me, waiting to be unraveled. Childhood memories inform; they shape the seabed of our worldviews. Although attitudes are formed from experience, they can evolve.

Maps

When traveling in Ecuador, we visit an equatorial monument in the countryside near Quito. A woman in indigenous dress is selling maps of the world. As I unfold its quadrants, I see that the equator is shown circling the globe north to south, instead of its usual horizontality around the circumference. The map key states: "The Equator is the line that unites the two hemispheres into one World. That is why Equator means 'equalizer: the line of balance, of equilibrium, of unity.'" I remember the blue globe of planet Earth on my desk at home, which I spin in search of place-names of particular countries. The grid of longitude and latitude has always shaped my worldview, but it no longer seems iconic. This makes me smile. It's bold to create a map that changes how we see. North, south, east, west: it's all one!

Inhabiting Betweenness

My Turkish colleague Nükhet refers passionately to the dichotomy between East and West. I feel confused at her depiction. What does East/West mean today, and what are my relevant assumptions? Her recent book, published in English and Turkish, tells the dramatic story of her grandfather's transition from the Ottoman Empire to what we now call Turkey. Spanning four decades, photos document radical identity shifts from turban, to fez, to European fedora, to bare head with mustache. As a doctor, he wrote books in Arabic, then French, then Turkish, communicating essential views. The choice was stark: Do you stay with your Eastern ways, or become Westernized, releasing lineage? I realize that she embodies that dual heritage, inhabiting betweenness, whereas for me it's merely an abstract idea.

an energized mindset and approach to your day. You'll be surprised how much impact attitudes have on your "happy" serotonin levels and your ability to take in new experiences and to cultivate relationships.

TO DO

NOTICING BREATH

5 minutes

Breath is the link between unconscious and conscious aspects of bodily knowing; all emotions and thoughts register in your breath.

Seated or standing, eyes closed:

- Bring your awareness to your breath.
- You may feel the touch of air on your lips, the movement of your ribs, or your belly moving as you breathe. Just notice any sensations of breath without doing anything about them.
- If your mind wanders, bring it back to this essential breath exchange between your body and the environment around you—taking in and releasing back out, expanding and condensing.
- Remember, there is no muscle of exhale. The out-breath is relaxation.
- Slowly open your eyes. Notice if you can maintain awareness of breath with your eyes open.

TO WRITE

WHAT DO I CARE ABOUT AND WHY?

10 minutes

Attitudes shape our actions and interactions. Write a paragraph addressing the questions: What do I care about, and how does that affect my work? What gets me out of bed in the morning and motivates my day? Write faster than you can think, writing to discover.

Underlying Patterns: Biology

Our family of origin extends back in time to the planet itself. Where we place our time-markers affects how we view the world. If our awareness is stuck in the atrocities of a particular war, the family feuds in our local community, or the spousal fight last night, our communication range is shaped by that lens. Yet we are impacted by cosmological, geological, and biological dimensions as well as historical, cultural, and familial ones. As we notice how personal story and scientific perspectives interact, we can appreciate complexity rather than be overwhelmed—acknowledging our heritage.

Overlapping edge zones between biology and biography have often been studied in psychology: the continuum between "nature and nurture" is discussed and debated.[1] It's a "yes–and" phenomenon. What's innate in our biological inheritance and "birthright" can be aided or restricted by context. History lives on in our DNA, cell structure, and neurochemical programming (how we are patterned to respond based on heritage) and is reflected in physical, mental, and emotional characteristics. Yet our context, will, and motivating desires play important roles along with moral and social imperatives.

The good news is that neuroplasticity exists. We can change behavior patterns and belief systems. The links between "wiring and firing" in the nerves create, quite simply, how we behave in everyday life. But they are rooted in more ancient, silent-level pathways of embodied knowledge. We can recognize and shift our inherited and conditioned patterns through skillful work with our own natures. The process starts with understanding and appreciating our embodied selves.

Considering the evolutionary story that underlies human capabilities, we can recognize that we are one family or species (*Homo sapiens*) and that we share this planet Earth as home. Being able to change lenses and timescales offers fresh possibilities for broadening communication across borders and perceived boundaries. For example, we can look at any situation through the lens of difference, which is important in keeping and valuing diversity on the planet—biological and human. We can also look through the lens of commonality: what we share with other humans and with biological systems. Rather than feel alienated or alone as individuals, we can recognize affiliation with multiple species and with our human kin across generations.

FACTS

In a simplified overview of the evolutionary timeline, based on the fossil record and close observation, we can look at a few key developments on the trajectory to our contemporary human structure and behavior. Life on

We are basically fluid creatures that have arrived on land. — Emilie Conrad, from "Letter from Emilie"

Guts

On the farm as a child, I had seen/touched/smelled "guts." The chicken I fed in the morning was often beheaded by noon (a hatchet and a chopping block), hopped headless around the yard to drain the blood, and ended up on the plucking table in our basement. There I had helped my father eviscerate "innards" from the chicken's body before my mother cooked the meat for dinner. The feel and shape of hard heart, slimy intestines, tough gizzard were familiar. So when Mr. Barlow, our charismatic seventh-grade science teacher, had us study the *inside* of the human body, biology became personal! The two-dimensional drawing of the human digestive system in our science textbook—forbidden knowledge in my Southern school—linked what was inside that bird to the inside of me: the rumblings and grumblings of my stomach and the escalating beating of my heart. Mr. Barlow was a rebel and believed we should know how things worked—including our own bodies. This was the first day of my future career, teaching (and appreciating) the see/feel nature of human anatomy.

Jennifer Nugent in *another piece apart* (2018)
Photograph by Ben McKeown

Earth began with single-celled organisms in a fluid oceanic matrix around 3.5 billion years ago. A semipermeable selective membrane both separated and connected the internal contents of each cell with the external environment. The three qualities of a cell, the ability to metabolize, reproduce, and respond to stimuli, are still inherent in each of the thirty-seven trillion cells that constitute our human bodies.[2]

From the oceanic matrix multiple body forms evolved. These forms included the asymmetry of the first living cell in a shallow tidal pool; the sharing of resources in multi-celled organisms, such as sponges or sea coral; the vessel-shaped digestive cavity of sea squirts attached to the ocean floor; the central organization of radial symmetry in starfish; and the bilateral symmetry of fish like the highly successful shark, with many permutations in between. The oceanic heritage of our species is still present in elements of our structure, such as the hollow tube of our digestive system and our segmented spine. Through all of these changes we retained our liquid core; the amniotic fluid in the womb and our blood match the saline content of the sea. Our skin keeps us from drying up. In fact, a human can be described as a bag of water walking around on land.[3] Although we may think of ourselves as solid and fixed, with a body that is over 70 percent water, fluidity is inherent.

As our predecessors washed up on shores and became land creatures, gravity and inertia placed new demands on previously successful forms. The skin became the mediator between the fluid interior and the air-exposed exterior, modulating the exchange of fluids and nutrients and maintaining a range of temperature suitable for life. What had been a mobile spine was now also used for stability, as the primitive tail and fins differentiated into legs for effective locomotion. The head lifted away from the earth to heighten the effectiveness of seeing, hearing, smelling, and tasting.

Across the evolutionary timeline, our land ancestors diversified from belly-slithering, to four-footed, to two-footed creatures. Eventually in the bipedal stance, hands were free for carrying, manipulating, grasping, and using tools. Hand and visual acuity required coordination, learning, and memory, and increased demands on the nervous system enlarged brain size. This included exceptional development of the prefrontal cortex—the front brain, which enhanced human capacities to communicate verbally, plan actions, and imagine outcomes. The older brain—the back brain—retained life-sustaining capabilities that every human infant is born with, including innate abilities for breathing, sleeping, waking, eating, excreting, and developing deep emotional bonds.

The primary characteristic of *Homo sapiens* (Latin for "wise man"), or modern humans, is this increased capacity of the brain. As humans evolved, all of our systems developed simultaneously in complexity. Our skeletal and muscular changes and physiological capabilities were matched by our capacity for three-dimensional thought in the past, present, and future. Humans are able to reflect on where we have been, contemplate and experience the present moment, and plan where we are going. The capacity for reflection, planning, and manipulation of our environment brings the responsibility of choice. We are responsible for what we create and for how we choose to inhabit that creation.

GRAVITY

Humans develop in the gravitational field. The body is shaped by and intimately attuned to gravity for basic survival responses—our potential for action and interaction. Recognizing and supporting orientation to ground and space creates the conditions for effective engagement and communication. Many reflexes and movement patterns are encoded in the body. With a newborn baby, *bonding* with the Earth underlies all other developmental responses. Healthy babies bond with gravity by releasing weight to be held, with air on the first breath, and with mother by touch and nourishment in feeding/suckling. We humans have to release our weight down to the ground in order to lift the head up. We have to feel support to push away, to initiate movement, and to reach toward "other" or pull "other" toward ourselves.

Condensing and *expanding* are considered basic cellular movements in the human body: condensing establishes ground, connection to the earth and self; expanding establishes spatial reach, connection to air and space.[4]

The Power to Kill

I can kill someone. It is not a mental thought but visceral knowledge about threat. When the adult son of a family I was staying with one summer gained my trust then threatened me sexually, if I'd had a knife, I would have killed him. I felt heat and rage surge, taking over any rational thought. Instead, I locked myself in my room—protecting us both. But the inner knowledge that I could kill is impactful. It's not just "others" who have that capacity; it's fundamental in each of us if the situation calls it forth. That's Kali, the Hindu Goddess, both creator of life and destroyer of evil forces. Generally, my outer rage comes not to protect myself but to defend someone else. I'm conflict averse, but when I saw a woman throw a broken bottle out of her car window at a child, I charged after her, yelling and challenging her action. We each have "the line" we draw between action and inaction. Just because you can do something, doesn't mean you should.

EVOLUTIONARY TIMELINE

13 bya	"Big Bang" (universe expanding)
4.6 bya	Earth forming
3.8 bya	First water
3.4 bya	First life, photosynthetic bacteria (single cell)
1 bya	First sexual reproduction
750 mya	Multicelled organisms (cell colony)
500 mya	Hydra and starfish (radial symmetry)
400 mya	Bony fish, first land plants and animals (bilateral symmetry)
350 mya	Amphibians and insects radiate; coniferous trees
280 mya	Reptiles radiate
250–200 mya	Pangaea and Panthalassa (single continent and surrounding ocean)
225–65 mya	Dinosaurs
180 mya	First small mammals and birds appear (reptiles rule)
135 mya	Flowering plants widespread
65 mya	Continents in their current positions
50 mya	Early primates radiate
22 mya	Early apes
5 mya	Ape/human split
3.7 mya	*Homo erectus* (bipedal alignment)
1.8 mya	First stone tools
75,000 ya	*Homo sapiens*

Dates indicate longer, overlapping periods of time; they also change with new findings. Key: bya, billion years ago; mya, million years ago; ya, years ago.

Macrocystis pyrifera, sporophyll blades
Art print by Josie Iselin
www.josieiselin.com

The rhythm of yielding and reaching, internal connectivity and outward expression, is the continuum we negotiate as human beings, basic to our existence and present in our communication patterns. We can feel it in ourselves as we relax into sleep and wake to meet outer challenges.

SOCIABILITY

Mirror neurons are part of the human sociability system. Mirror neurons respond to actions that we observe in others, often imitating them through mimicry and empathizing. When a mother smiles, the baby responds in mutuality.[5] *Entrainment* is a phenomenon in which humans change breath and heart rates to match each other's patterns and conserve energy. In communication, there's a pull toward synchronicity; we entrain with others in order to empathize, to work together, to live in social groups. *Resonance* is a state of feeling in tune with a person or place involving both mirror neurons and entrainment. Accompanied by easeful breath rate, our body rhythms are in sync, and there is a fundamental comfort or safety level that permeates interactions. Resonance allows affinity in communication and enhances sociability—with or without words. Nonverbal forms such as music and dance cross linguistic barriers and encourage easeful entrainment and resonance.

Connections

The focus today is on experiencing fluidity. Imagine your whole self as the fluid being that you are. Inviting flow at various times in your day can refresh and restore your energy. If you feel good, you have more to give those around you. A day spent in flow may surprise those who are used to your more edgy or restrained demeanor. Enjoy—your fluid self is your birthright, present and available. Flowing water can be powerful as well as calm; it breaks down rock. By finding fluidity even in challenging situations where tension is expected, you are prepared and rehearsed for resilience and recovery, appreciating the relevance of your ocean ancestors to your own fluid nature.

TO DO

POURING THE FLUID BODY (CARYN MCHOSE) ◉
5 to 10 minutes

Rolling and pouring returns us to cellular awareness: no top, no bottom, no gravity. Enjoy!

Lying on the floor or seated, eyes closed:

- Imagine you are a single cell, floating in the ocean with fluid inside, suspended in fluid outside. Your skin is a semipermeable membrane, selecting what flows in, what flows out.
- Pour your fluid contents in any direction. Imagine you are totally suspended in a fluid, warm sea, moving and being moved with the tides.

- Now, bring your attention to your skin, the selectively permeable membrane. Move with your focus on the skin, the outer membrane. You are touching and being touched simultaneously through the skin.
- Keep pouring and rolling while bringing your attention to the fluid contents inside the skin. Pour the fluid into particular body areas to initiate movement like an amoeba with a temporary protrusion of the protoplasm, a pseudopodium, that serves in locomotion and food gathering.
- Move your body slowly so that you can perceive sensations. In this asymmetrical movement there is no up, no down, no head or tail, no right or wrong way to move; enjoy the sensations of wholeness and disorientation.
- Pause, noticing what is occurring. This state of nonjudgmental wholeness is called "open attention," simply noticing sensations, emotions, thoughts, and images as they arise.[6]
- Slowly add vision, remaining aware of sensations. What is your experience?

TO WRITE

FAMILY PET
10 minutes

Animals are our ancestors and companions; they inform our understanding of the natural world—including our human bodies. Write about a family pet. If you didn't have one, write about your relationship to any animal at a young age. What did you learn from your animal relatives, and how does that affect your ways of communicating now?

Diving In
Diving underwater, I can almost hear the sizzle of my brain cooling. It is quiet in the underwater depths. My body feels sleek, undulates like a fish with a mobile tail. Reminded of "The Little Mermaid" story by Hans Christian Anderson from my Danish ancestry, I feel both land and water as home.

Cicada, a French alpine goat
Drawing by Louisa Conrad
Big Picture Farm, Townshend, Vermont[7]

Cameron McKinney
Artistic director of Kizuna Dance
Photograph by Peter Yesley

Underlying Patterns: Biography

Day 4

Our personal life story begins in the womb and continues to this day. Reflecting on this heritage helps us know ourselves better and communicate more effectively with others. Ancestral lineage, including place origins of relatives and our present-day cultural and intercultural contexts, plays a role. Gender, sexual orientation, race, class, politics, culture, profession, family, and daily life all inform. Recognizing that the egg cell that made you was formed inside your mother's fetus while she was inside your grandmother's womb underlines the impact of ancestral lineage.[1] And we know that physical environments, including stress levels and nutrition, register in the developing body. History is encoded in our body-minds and is constantly unfolding.

In the early months of life, the threshold between biology and biography merges. All sounds are equally possible to hear at birth, for example, but the child neurologically begins to preference the sounds related to the language and vocal characteristics of their parents at about five months. Human development patterns (such as yield and push, reach and pull) are encoded in all bodies for survival, but situations can limit or facilitate their expression. For example, studies show that babies in orphanages who were isolated for hours in cribs, unable to move through the early developmental patterns, often have long-lasting, movement-coordination challenges.[2]

Every person has a cultural inheritance that registers in the body. That inheritance can involve degrees of trauma and abuse over the past centuries of wars, feudalism, and colonization. Context affects embodiment. How we inhabit our bodies now is a locus for embodied social change. Recognizing one's personal story and what led to their conditioning, regardless of current social status, can help unravel outworn patterns—physical and psychological. Healing ourselves can involve healing relationships with our ancestors as we create new stories about how best to live together on Earth.

SPECIFICS

We all migrated from somewhere. Distinct landscapes, culturescapes, and belief systems have shaped the characteristics of individuals and groups. Some ancestors thrived while others faded in our lineage. Political upheavals, natural disasters, and migration patterns have marked certain lives and left others unscathed. Is your family lineage intact and located in place? Were southern climates part of your skin color and UV-protecting melanin levels? Is your sense of timing and time related to culture and climate? Did your relatives flee war or famines and emigrate? Did near-starvation inform eating and food-sharing patterns? Was enslavement and torture part of your lineage? Are you tuned more to nature or to people for comfort? Frameworks for understanding the body and psyche can be helpful

Allowing ourselves and others the multi-self self . . . — Julia Alvarez, discussing Afterlife

Signs

My grandmother, I am told, wrote the horoscopes for the local paper. She would sit down on Sunday afternoon and create them at the kitchen table. I think about this when someone asks my astrological sign. When I say Leo, there's generally a knowing nod—bossy and organized! But is there more? Today near Leeds, England, we stroll through the Yorkshire Sculpture Park where Ai Weiwei's "Circle of Animals" depicts twelve creatures of the Chinese zodiac arranged in a circle on tall poles. Rat, ox, tiger, rabbit, dragon, snake, horse, sheep, monkey, rooster, dog, and pig. Crowds gather at their bases, identifying their "sign." Mine is a rat; I stare longingly at the expressive dragon across the way and try to reframe the creature a rat represents in me. Rat is the first in the twelve-year cycle of the Chinese zodiac (including 1948—my birth year). "Though people consider the rat not adorable, and it even makes its way into derogatory language, it ranks first on the Chinese zodiac signs. It has characteristics of an animal with spirit, wit, alertness, delicacy, flexibility and vitality."[3] Surprise!

Thinking/Sensate

In Jungian typology, my dream therapist and I decide that I am a "thinking, sensate." These are my first and second "functions." I think my way into dancing and am an organizer and planner. Intuition, my "third function," is less reliable, even lazy, although when it comes on board it can be quite profound (go there, now!). Way at the bottom, feelings, my fourth function, are hard to access—and emotional values pose quite a challenge. It has been helpful to know this model through the years, making sense of behavior with less self-criticism and more discernment. Supposedly, in this second half of life the functions invert: I'm now an "intuitive feeler" and more extroverted. No wonder I often feel awkward in situations that used to be easy. It's a whole new view.

in understanding who you are in relation to history and are essential for reflecting on communication patterns inherited from your family, friends, and cultural or historical context.

HISTORICAL PERSPECTIVES

Various models for understanding psychological integrity have been developed throughout human history to enhance individual wholeness while living and communicating in social groups. Astrological signs based on the alignment of stars and planets on the day of your birth are perhaps the most ancient, universally utilized indicators of personal characteristics. Understanding astrology—the ever-changing pattern of the planets and stars in relation to human experience—helps link the patterns of one's life to the patterns of the cosmos. The goal is not to dismiss personal responsibility but to use the underlying neutrality of astronomy and mathematics to locate oneself within the larger context of the unfolding mysteries of life.

One of the more recent theoretical frameworks is Jungian typology, developed by Carl G. Jung, a Swiss psychiatrist and psychoanalyst who developed analytical psychology in the mid-1900s. Through his own inner work, Jung discovered, studied, and articulated his theory of four psychological types: thinking, feeling, intuitive, and sensate. Understanding typologies helps individuals recognize personal orientations and patterns to bring life to less developed parts of themselves that remain in shadow. Throughout a lifetime of personal individuation—one's integrative work toward wholeness—Jung also identified the tendencies that lead toward an introvert or extrovert personality.[4]

Psychological models and theories have developed into contemporary personality "type tests" such as the Myers-Briggs tests for compatibility, the DISC profile (acronym for dominance, influence, steadiness, and conscientiousness), and the widely used Enneagram System (which identifies nine distinct personality types and their overlapping qualities). The focus of these frameworks is on understanding and strengthening the less-known aspects of yourself that would help you engage and collaborate more broadly and respectfully with others. Discerning one's strengths for optimal living and performance can reduce stress and enhance well-being within the complexities of our contemporary, fast-paced world.

The challenge of working within any model is avoiding fixation on a single way of viewing oneself and the world. Effectiveness involves using a specific framework to support and expand your horizons. It can also grow less-familiar qualities within yourself as part of the lifelong inquiry of self-development, possibly allowing you to overcome a limited way of experiencing life and communication in a global context. By becoming less judgmental and more encouraging of the different components of each person's psyche, you can support others' strengths and help them explore new possibilities. As a teacher, mentor, or leader, cultivating awareness of multiple potentialities within oneself and others invites longevity of relationships and dynamic projects.

LIMITATIONS AND POSSIBILITIES

We can recognize what restricts or enhances personal growth. *Fear* can stop interactions or move things forward; and fear of fear can immobilize us. But when partnered with *curiosity* within a safe context, fear responses can be channeled to useful ends. Excitement is part of activation: we can use this energy to move forward when appropriate support is at hand. Understanding the roles of the amygdala and vagus nerve in fear responses can help us move beyond inherited or developed reactivity into present-moment choices toward wellness and cultural cohesion (see Day 9).

Trauma recovery involves the whole person. Dr. Stephen Porges, whose work in neurophysiology on polyvagal theory now informs many treatment protocols, reminds us that the human nervous system is "one nervous system": there is no mind-body split. Becoming co-regulators with our bodies is an emergent process—occurring moment by moment. Mind can change organs, and organs can change mind through processes such as heart rate variability. There is continuous interaction, with complex feedback loops for monitoring internal balance. This includes the role of the vagal nerve in the multilayered fight-or-flight, freeze-or-faint, and tend-and-befriend responses of the autonomic nervous system, with implications for trauma treatment and recovery.[5] He affirms that polyvagal-informed practices are part of a toolkit offering a journey of knowledge and health as we take responsibility for the impact our physiological state has on others.

CULTURE

Every culture has its distinct qualities, and multiple subcultures affecting interactions exist in any community or organization. Most of us live in the realm of interculturality, including intercultural (between nationalities) and intracultural (within one culture) dimensions and requirements. Friends and colleagues enter our lives from different countries as well as from distinct cultural heritages.

WORKING TOGETHER

Trust is an important theme in personal communication. Trusting and being trustworthy are not to be assumed. Trust allows collaboration between groups and individuals with minimal friction. Understanding the different potentialities and motivations within human beings—including psychological typologies and social and cultural values—lays the groundwork for trust. Multiple individuals and groups can intersect cohesively when there is a sense of reciprocity and shared values. With trust, we find ground under our feet, unspoken but agreed-upon goals are present, and time is utilized efficiently. When trust is not possible, has been violated, or its requirements have been unacknowledged, feelings can easily go astray. Nonverbal and emotional cues are misinterpreted through projection, negative expectations, or past associations.

Repair may be essential if it is possible to work within the bounds of a safe and mediated context. A sense of humor about individual diversity,

Discovery

Somatics author Don Johnson reflects that body-centered therapies not only sensitize you to be more nuanced in what you perceive and how you respond, they also strengthen you physically and mentally. At an Embodied Yoga Summit, he offers personal experience that, especially when partnered with martial arts practices, you develop physical endurance and resilience. You can defend yourself, resulting in confidence that allows you to be more generous to others, less fearful.

Four Views

David Brooks, in *The Social Animal*, asserts: "Trust is habitual reciprocity that becomes coated by emotion. It grows when two people begin volleys of communication and cooperation and slowly learn they can rely on each other." In *Horizon*, Barry Lopez writes: "We are the darkness, as we are, too, the light." In *My Grandmother's Hands*, Resmaa Menakem reflects: "All adults need to learn how to soothe and anchor themselves rather than expect or demand that others soothe them. And all adults need to heal and grow up." Dan Siegel adds, in a talk about *Mindsight*: "Learning to be open to many layers of communication is a fundamental part of getting to know another person's life."[6]

Western Lady Dream Platter (2018)
by Mara Superior
High-fired porcelain, ceramic oxides, underglaze, glaze, gold luster, 12.25 in. × 15.5 in. × 1 in.

personality proclivities, and stylistic choices creates ease. Good humor and play can enhance creative thinking and allow integrated, oriented, and focused exchange. Kindness is involved. Somatic investigations support effective communication skills. They offer direct experience of the intelligent moving body. We are learning to restore connection to the body as subject, not object, and to respect ourselves as well as others in the context of history as well as the present moment.

Identifying with another person's narrative supports empathy, but if we merge completely, we lose self-awareness and stability. There is a neutral distance required in embodied communication; grounding and spatial awareness allow whole-body orientation. From partner dances like salsa or tango, you can learn about "leading by following." Responsiveness is essential; there's a constant give and take between sensory impression and sensory expression—what you feel and how you respond. Maintaining the integrity of oneself in the presence of another is an ongoing process.

Movement practices remind us of our inherent resilience and adaptability; there is not just one way to be in the world. Shifting through different states of attention, we practice the capacity to respond to and accommodate changes in external and internal environments. Movement dissipates heat, hydrates tissue, and restores wellness mentally and physically from inside: there's a brain-body balance creating resonant flow.

Connections

Self-knowledge at physical and psychological levels allows you to have somewhere to come from as you meet "other." The invitation today is to focus on the ways that history is encoded in your body-mind. As you become aware of your personal history, including the stories you tell yourself about yourself, you can clear unnecessary or tangled pathways and be more present in relationships. This might involve jotting down a few memories or experiences that surface during your day. The dialogue between your past, the present moment, and your anticipation of the future intertwines in each word and gesture you choose all day long.

TO DO

INNER OBSERVER

5 minutes

> *A non-judgmental and supportive inner observer can help balance internal awareness with outer demands. You can invoke your inner observer several times during the day to "check in" with the body rather than "check out."*

Seated or lying on the floor, eyes closed:

- Invite a supportive inner observer into your body as a witness.
- Allow this observer to enter with your breath and travel slowly through your body, shining a light on inner parts.

- What does your inner observer notice? Sensations and sounds, shapes and colors, bones and organs, memories and images, expectations or resistance? Just be aware without judgment.
- Now, slowly open your eyes.
- Invite a word or words to describe your experience. Be simple, noticing how language emerges from the body. Speak this word or phrase out loud: "Cold." "Peaceful." "Unsettled." "Trembling." Let it be your own word.

TO WRITE

BODYSTORY
5 minutes/2 hours

Every physical body is full of stories. Choose one body part and write about it for five minutes (example: your foot, elbow, chin). Surprise yourself—you never know what is stored in an area—there may be several memories. Then give yourself two hours (or two weeks) and write your full bodystory—the history of your body from birth to now. Include:

- The story of your birth (pre-birth if possible; the health and activities of your mother affect life in the womb)
- Your earliest movement memory (earliest kinesthetic sensation you can remember; examples: being rocked, learning to swim, bouncing on your parent's knee, riding a bicycle)
- Training techniques (sports, dance, gymnastics, musical instruments)
- Environment where you lived (mountains, plains, forests, oceans all affect how you move, how you perceive)
- Comments you heard about yourself that shaped your body image ("Oh, what a cute chubby child! Stand up straight! He's going to be tall like his dad. Children are to be seen and not heard.")
- Attitudes toward sensuality, sexuality, gender images
- Injuries, illnesses, operations
- Nutrition, relationship to body weight, strength, flexibility
- Fears, threats, real or imagined
- Ancestral lineage: where your ancestors lived and their conditions

Put your bodystory away for a few days, and then reread with fresh eyes, appreciating all that has shaped your life.

Shaking

As I finish writing this chapter on underlying patterns, I am shaking. Sitting down and consuming a big breakfast of comfort foods helps: eggs with cheddar cheese, fresh-picked wild blueberries and yogurt, cornmeal mush and Vermont maple syrup. You can't dig around in biography and not stir up feelings, I think to myself. We all have parts of ourselves tucked away that don't want to be revealed. And there's no reason to write a book unless you excavate some of those places. Shaking says: "Hello. Yes, you are in dangerous terrain. Take a break. And keep going."

Day 5

Perception

We perceive the world as we are. Our senses are affected by the state of our body and the quality of our attention. At an internal level, this can be the alchemy of food, sleep, and hormones affecting the chemistry of our bodies. At an external level, people, place, and weather patterns affect interactions. Throughout, there are values and hierarchies of conscious attention: what we care about and what we have been told is important. Interpretation of sensory data is based on our past. Stimuli we have experienced before get through perceptual gateways first; habits inform.

Perception is how we make sense of the world. Sense organs throughout the body tell us everything we know about ourselves and the world. We all have the same sensory organs, which are similar, but the ways we perceive the world are distinct. No two people are perceiving the same thing in the same way at the same time. This is an important baseline for understanding the process of communication and avoiding the pitfalls of miscommunication and conflict. Billions of sensory receptors throughout the body constantly feed us signals in a process we call *sensing*. Our ability to

The Belle Cabinets (2013)
by Kristina Madsen
Maple (painted, carved, and
gessoed) with wenge base
Photograph by
Stephen Petegorsky

organize and decipher these signals is called *interpretation*. Our *response*, which may be to not respond, completes the process.[1]

We are active participants in these three phases of *perceiving*. We can heighten awareness of *sensing*, broaden patterns of *interpreting*, and encourage new pathways of *responding*. Change in habits occurs in the "pause" between sensing and interpreting; we can take in more sensory data. Change is also readily available in the "space" between interpretation and response, so we can allow more time for fresh connections. For example, you see a man approaching you in disheveled clothing and quickly pass him by and then find that he is the director of the funding organization with whom you are going to meet. As you pause for more sensory cues, you remain open to possibilities. Understanding this perceptual process can help us *act* from the actual sensory information available at the moment rather than *react* from habit, fear, or uninformed associations. Awareness enhances our ability to respond—our responsibility. Recognizing the impact of perception may be the most essential component of communication.

FACTS

We construct our ideas about reality through the specificity of human perception. The nervous system responds to only a selective range of wavelengths, vibrations, and other stimuli. For example, human eyes utilize a particular spectrum of light for vision; it is the same spectrum of light that plants use for photosynthesis, but it is not the full spectrum. A bee employs light in the ultraviolet range for navigation—a type of light that our eyes don't recognize. Human ears are equipped to respond to a particular range of vibrational frequencies. These are the same frequencies employed by birds, which may partially account for human fondness for their songs, but it's not the full range of sound. Whales and bats use higher frequencies. There is more to life on Earth than we are able to perceive.[2]

Perception is culture-bound, shaped by education, religious and political systems, language, and arts. For example, instrumental music that might evoke a trance state in Morocco could sound irritating or chaotic to an uninitiated foreign listener. American educational systems are generally oriented toward visual-auditory learning. Literacy in the areas of movement, touch, and visceral response (including emotion) is often ignored, limiting our capacity for subtlety of expression, communication, and problem solving. Senses respond to particular landscapes, smells, tactile stimulation, gestures and sounds; our interpretation of this input becomes our experience, based largely on cultural conditioning.

Perception is also personally selective, affected by genes, family values, previous experience, and the individual's current state of alertness. Beyond very basic inborn survival patterns, we learn to focus our senses. No two people perceive the same reality. Ask a group of thirty engaged in the same activity what they are noticing, and each will have a very different view. In a conversation, unique and sometimes contradictory interpretations will occur. We each have perceptual strengths and biases. For example, one

What Do You Notice?
I was trained to see. My father was a watercolor painter, and all three of his daughters have college degrees in visual art. As children, piling out of the car on vacations, we would line up on a rock and "look," silently. I thought this was how everyone perceived the world. Part of our young training was to view landscape or cityscape compositionally, noticing foreground, middle ground, and background and choosing where to emphasize focus. When I traveled with a friend's family, she and her brothers scattered, cavorted, and called to each other as part of arriving—it was a different lens for experiencing a place.

Noticing What Works

Adam and Aaron direct a greenhouse project in Peru called the Andean Alliance for Sustainable Development (AASD). They Skype with our class to talk about international development projects. Their practice: "We look for what is working in a community," they tell us. "Aid agencies often go in and want to 'fix something.' They generally have a limited time span, specific budget, and have to accomplish a project to keep their internal funding. And then they leave." Instead, Adam and Aaron work with a community-driven model; they visit communities to see what they are doing *right* and amplify those resources. "People may be disappointed initially; they are used to aid organizations giving them money to fix things. But then, they take satisfaction and ownership over their own project. We assist, but do not dominate, so projects can have both life and longevity."[3]

person in a classroom might notice how hot the space feels, while another is distracted by the volume of the speaker's voice or the visual details of clothing and gesture.

Expectations affect perception: we see what we expect to see. For example, if you think someone doesn't like you, you may interpret visual cues as negative to support your pre-formed expectation. In some cases, broadening expectations may allow us to detect more subtle signals. When we look at each other, most of us see solid bodies. But if we learn that electromagnetic fields exist around our forms and are measurable in a laboratory, we might experience the body more expansively. We are used to feeling a personal blood pulse, but if we learn that there is also a cranio-sacral pulse, slower than the heartbeat (six to twelve cycles, instead of the sixty to seventy heartbeats per minute), we might then begin to perceive (tune in to) the craniosacral rhythm. Our expectations about perception can limit or expand our perceptual range.

In this three-part model, which includes the human perceptual system, cultural layers, and personal patterns, our perceptual range narrows. However, through information and experiential exercises, we can reinhabit our full potential. Perception itself is highly plastic; there is no objective reality. As we begin to recognize and change the filters of association and categorization through which incoming signals are screened, our perceptual conditioning and habits can expand.

Perception is whole-body. There are four generally accepted categories of sensory receptors. The *interoceptors*, found mainly in the organs and blood vessels, are responsible for monitoring the inner workings of the body, such as blood chemistry, heartbeat, and digestion. (If you've just had coffee, you may feel a caffeine "high"; if you need food, a blood sugar "low.") The *exteroceptors*, found in the skin and connective tissue, are responsible for monitoring the outer environment through touch, including several kinds of sensations such as pressure, heat, cold, pain, and vibration. Sensation is specific: different types of exteroceptors have individual names (often based on the scientist who discovered them). *Pacinian corpuscles* tell of the deep pressure of your feet against the floor, *Krause corpuscles* register cold, and *free nerve endings* register the light touch of cloth brushing against your skin. We can use naming not to distance and dismiss but to consider these essential sensory detectors in our bodies as family, familiar.

The *proprioceptors*, found in the joints, ligaments, tendons, fascia, muscles, and the inner ear, are cumulatively responsible for registering movement and body position in space. If you shift how you're sitting, your proprioceptors tell you that you are moving and the resulting position. The *vestibular system* located in the labyrinth of each inner ear gives us our sense of balance. Equilibrium is important in survival as we negotiate our way around changing terrain. The *special senses* include sight, sound, smell, taste, and touch.

Of the special senses in humans, *smell* (olfaction) is the oldest, responding to chemicals in the air. Smell registers directly in the olfactory bulb (part of the limbic system, the emotional-relational brain) and is

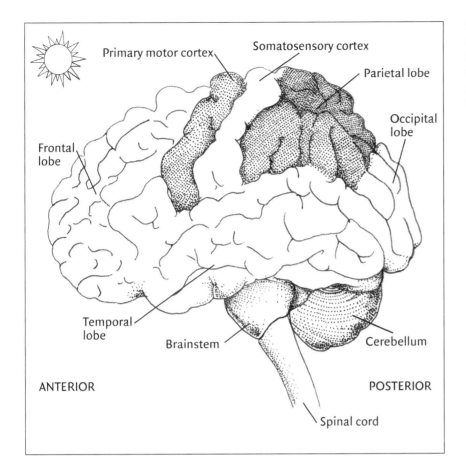

Primary motor cortex

Somatosensory cortex

Parietal lobe

Occipital lobe

Frontal lobe

Temporal lobe

Brainstem

Cerebellum

ANTERIOR

POSTERIOR

Spinal cord

Somatosensory cortex of left cerebral hemisphere (exterior view)
From an MRI of a twenty-four-year-old female dancer
Anatomical illustration
by Nancy Haver (2021)

closely linked to survival responses. Often interpreted below conscious awareness, smell affects emotional memory, sexuality, and "gut feelings." *Taste receptors* are also chemoreceptors. Located in the taste buds of the tongue and—to a lesser degree—the soft palate, these receptors register bitter, sour, salty, and sweet flavors in various combinations. The taste of foods we eat or reject is strongly influenced by smell and also by texture, temperature, visual appeal, and personal and cultural history. *Hearing* is constant—our ears are always open to signals in the environment. *Vision* is both the fastest and the slowest sense (see Day 21), and *touch*, the fifth category of special senses, includes all the other sensory modes—exteroception, interoception, and proprioception.

In the big view, all senses are one. Perception is cumulative and generally simultaneous; rarely does one part of the sensory system act alone. Nearly all sensory signals go first to a relay station in the *thalamus*, a central structure in the brain, for integration with other sensory input and then to primary sensory areas of the cortex, including the *somatosensory cortex* for interpretation and the *motor cortex* for response. For example, dimensional interpretation of a visual image is cued by information from touch, smell, taste, sound, sight, movement, and visceral activity. When you are watching two people talking, you are also hearing their voices and background sounds, experiencing your body, and responding viscerally to the experience.

Updates

In his book *When the Body Says No,* addiction expert Gabor Maté reminds us: "The physiology of stress eats away at our bodies not because it has outlived its usefulness but because we may no longer have the competence to recognize its signals."

Perceptual selectivity is essential. If we were to be aware of all the sensory information coming into our bodies at any moment, we would be overwhelmed. Thus, intention affects attention, shaping what we perceive. For example, you might be drawn to the sound of the cell phone ringing while focusing on reading. The nervous system prioritizes change, which could mean threat or opportunity. What was background suddenly becomes foreground, influenced by our general state of health, alertness, and motivation.

Perceptual gateways are junctures where sensory information is filtered. This sensory discernment necessarily inhibits and deletes large quantities of stimuli and determines which go on to other brain centers for interpretation and which register at subthreshold levels. In neurological language, we have to open the attentional gates for stimuli to pass through and be made conscious, and the most frequently used pathways are most easily accessed. As we become aware of perceptual habits, we can make a practice of inviting new information, opening new opportunities for response.

Projection also affects perception. Projection is the tendency to place our own feelings, desires, and fears outside ourselves and assign them to other people, places, and objects. Some psychologists consider that as much as 90 percent of what we perceive "out there" is actually "in here." Often it is a seldom-noticed aspect of ourselves that is projected into the environment and affecting communication. For example, if you see a stranger and decide she is angry, you can be projecting your anger onto that person without acknowledging that it is possibly partly internal.

Othering is a component of projection that occurs on cultural or personal levels, where difference is construed as a threat.[4] Projecting our own unexplored nature outward can be an excuse for violence, war, and environmental destruction. Us/them categories are limiting and potentially dangerous; and group-based identities, while functional and satisfying the social need to belong, can also intentionally exclude or threaten. Yet projection is part of perception; our experience of the world around us is filtered through our inner life. As we become aware of the process of projection, we can engage its useful dimensions. Recognizing projection as an aspect of perception, we become aware of our personal, potent role in constructing our view of the world.

Numbing, dissociation, and *rigidity* are prevalent in our era. We are bombarded by sensory stimuli, much of which has emotional content beyond our capacity for response. Our daily dose of media news and images about the escalating environmental or political crises can require that we protect ourselves by shutting down sensory pathways. For example, as we look at a color photograph in *National Geographic* showing a distraught mother holding her dismembered child after a bombing, the pain is overwhelming. Or when we confront an image of a polar bear stranded on an ice flow in Alaska due to rising sea temperatures, it hurts to know these animal relatives are going extinct from global warming. We do not want to see. It's sometimes easier to just shut down, feel nothing, and keep moving.

Although "shutting down" is a healthy, even lifesaving response in specific situations, it is limiting, even dangerous as a constant state.

As we become less receptive, we miss the delight and sensuality that come from being engaged with life. However, once we can recognize a continuum of perceptual receptivity from fully open to fully closed, we can choose the degree to which we need to protect ourselves for the moment, creating temporary but useful boundaries. At the level of the cell, the selectively permeable membrane monitors what enters and leaves, moment by moment, for balance and homeostasis. In our daily lives, receptivity varies in response to stimuli, allowing us to participate actively and with discernment in the process of sensing, interpreting, and responding.

SENSORY RECOVERY

To restore the depth of feeling that comes from a responsive sensory system, we can pause and feel. Sustaining our ability to be conscious of sensory input involves staying open to what is actually happening rather than deferring to our thoughts or judgments about what is happening. As we reflect and interpret, we can continue cycling back to the senses to see if what we think and imagine is real and if our potential responses are relative to this specific situation. The more refined our awareness of sensory input and interpretation, the more opportunity we have for reflecting both analytically and empathically when engaging with the world.

Connections

You might set your intention today to notice sensations, listening to the language of your body. Shining a light on different parts of your physical self and your environment through sensations can wake up hidden stories and resources. As you bring them into awareness, you might just make different choices toward amplifying your life goals. What best serves your energy for the moment at hand? Self-affirmations can be helpful. Try saying, "I am okay just as I am" a few times during your day and notice the effects. 👁

TO DO

FULL BODY SCAN
5 minutes

> *Closing your eyes during a body scan allows you to feel sensations more clearly. But you also want to feel safe; if hesitant, try both.*

Seated comfortably or lying down in constructive rest, eyes closed or soft gaze:

- Take a few deep breaths to settle your nervous system.
- Bring your awareness to the top of your head. Notice any sensation you feel on the top of your head. It might be a tingling, a vibration, an itch, or pain. It might be a feeling of pressure, heat or cold, the touch of air on your scalp. (There are nerve

endings called *exteroceptors* throughout your skin telling you about your connection to what's outside the body.)

- Continue to observe any sensation you feel on the top of your head. (The repetition of language helps focus attention.) If you feel nothing, just wait—your nerve endings will feel more sensitive as your body prioritizes attention to sensation.

- Bring awareness to your face and scalp. Observe any sensation without judgment; the task is to feel what is actually happening in your body without evaluating whether it is good or bad, pleasant or unpleasant.

- Move your mind's eye to your neck. Remember to give equal attention to any sensation that you feel on your neck—such as tingling, the touch of cloth on the skin, your hair as it brushes the surface.

- If thoughts come, just observe—as though they are clouds in the sky. You notice them and let them pass without attachment. Sometimes it helps to just say "thinking" and return your awareness to sensation.

- Continue to your right arm, the left arm. If your mind wanders, just bring it back to the present-moment awareness of sensations.

- Move your attention to the back surface of your body, the front surface of your body, the pelvis, the right hip and thigh, the right lower leg and foot, the left hip and thigh, the left lower leg and foot.

- Bring your awareness to the soles of your feet.

- Now, return to observing your breath as it moves through your nose and mouth and into the body. (Internal awareness registers through nerve endings called *interoceptors* in your organs and blood vessels.)

- Take a moment to stretch through your whole body; move any way that feels good. (Movement stimulates the *proprioceptors* in your muscles and joints.)

- Slowly open your eyes; allow yourself to remain aware of sensations as you include vision.

TO WRITE

THREE THINGS

15 minutes

If you feel overly self-absorbed or distracted during your day, noticing "three things" is a quick way to refresh and rebalance connection to the present moment. Begin where you are *or* take a short walk, noticing three things:

- Three things you see. (Example: I *see* two people standing, a rock, a writing notebook.)
- Three things you hear. (I *hear* ocean waves, an airplane, and my voice greeting a passerby.)
- Three things you feel. (I *feel* air on my skin, cloth touching my arm, and wet sand under my feet.)
- Pause, and wait in open attention, perceiving whatever captures your awareness. (5 minutes)
- Now, write about your experience. Which senses were strongest for you today? What senses were prioritized in your childhood? Was there any one person who influenced your perceptual learning? (10 minutes)

Detail: *The Belle Cabinets* (2013)
by Kristina Madsen
Maple (painted, carved, and gessoed)
with wenge base
Photograph by Stephen Petegorsky

II
FOUNDATIONAL SKILLS

Uplifted Ground (2015)
Sculpture by Michael Singer
Austin International Airport, Austin, Texas
Photograph by Jason Bregman

Arriving and Orientation

Arriving is a process. How do we arrive so that our whole self is present? What do we need to settle in and open to other people and to the uniqueness and potential delight of a new experience? Locating oneself in the present moment is key for survival. Arriving effectively, whether it is during travel, in a conversation, or in this moment of reading, requires the foundational security of knowing where we are physically. Cultivating presence is a moment-by-moment practice that includes specific stages, bringing the body's intelligence "on board" for effective decision-making and communication.

If we do not take time to give our deep, body-level intelligence the support it needs for arrival and orientation, we waste time through distraction. Can you feel your feet? Is your spine relaxed and free? Is your breath calm and deep? This three-step sequence, starting with your base, helps create the conditions in your nervous system for clear thinking and communicating.

FACTS

Gravity is the constant background for all movement on this planet. This includes the movement of breath and our ability to stand and face others. The body has inherited many sensory receptors in the hands, feet, head, and tail specifically cued for location. As we direct our attention to these portals or gateways (the tonic system), we receive a deep full breath, the breath of location.[1] Sometimes we need to rebuild this perception many times during the day, feeding substantial subcortical data into the nervous system to answer the question "Where am I?"

Described by French anatomist Hubert Godard as "the gravity response," the tonic system importantly cues pre-movement in the body—the readiness for action.[2] This foundational survival process establishes the edge of nervous system alertness: are you unsettled and overly excited, or relaxed and available? Is your body preparing to defend or simply to engage? Once the pre-movement is activated in the neuromuscular system, the outcome is largely determined. For example, if you are anxious and insecure in your base, your breath, voice, and alignment will communicate this lack of confidence, regardless of what you say.

Recognizing and supporting the tonic system creates a dialogue with the "potential for action" of the body. Orienting to ground, space, and breath through the tonic system enhances our relationship with ourselves and opens awareness toward cultural and environmental contexts. This helps to bring integrity to any situation: conscious orientation starts with establishing optimal pre-conditions in the body.

Sometimes it's necessary to go a long distance out of your way to come back a short distance correctly. — Edward Albee, from The Zoo Story

Creating the Conditions

Beginning a meeting or workshop session is important. My mentor Janet Adler once described the importance of first steps as being "impeccable in the beginning." These words stay with me as I arrive early to clear the room, add flowers, set the tone of the exchange. Otherwise, you spend the whole session just getting on track. What are you trying to evoke, and how best should you prepare—the space, the greetings, and the first words—to introduce intent?

STRATEGIES

Arriving efficiently involves maintaining backspace: your depth and dimensionality in encounters. (You aren't flat!) If you feel disoriented, try standing with your back against a wall. Wake up the nerve endings along your entire back body. When you step away, can you still maintain your backspace in the presence of other people and the challenges of interaction? Rather than losing yourself in a conversation or getting emotionally hijacked by fear or defensiveness, can you stay grounded and spacious while maintaining omnidirectional awareness?

Orientation involves familiarity with the physical environment, even if your focus is on people. Many cultures and individuals orient themselves by the local mountain and water source. In a Māori greeting in New Zealand, for example, you might introduce yourself by saying your mountain, your river, then your name and human lineage.[3] This puts you in geographical context—encompassing billions of years of history as well as ancestral heritage. Becoming familiar with place includes acknowledging seasonal shifts and weather patterns, offering practical knowledge that underlies human planting cycles and rituals. Take a walk and notice water, air, and soil quality, all of which impact health and social interactions.

Biology is also important. What are the animals—including the diversity of birds, fish, reptiles, and insects—of a particular locale? These relatives and their sounds animate the landscape. The local plants and animals provide food and company as well as spiritual connotations with links to more intuitive knowing. Skins are used for drums, clothing, and adornments. Traditional garments, household objects, and culinary specialties have evolved over generations. Scanning through the biological lens as you arrive supports connection to place-based cultural values.

Architecture has impact. There's nothing neutral about the room where you meet for a gathering. Sometimes, to arrive fully, you have to distinguish the peculiarities and unique characteristics of the site. Take a walk around the space, notice what's on the walls, clear assumptions, and invite the open gaze and wonder of direct experience. Space and place impact perceptual responses, including interactions and comfort levels in communication. How near do you stand to another person, where is the appropriate place to sit in a group, what do you intend with your presence?

Establishing a relationship to a new place may include orienting to cardinal directions: north, south, east, and west. Where does the sun rise (east), where does it set (west)? Taking time to build a connection between where you came from and where you are now can help to embody arrival. Sometimes placing an object, creating an altar, establishing a particular relationship to the space and place where you are living creates ease in the nervous system so you can open more fully to new dimensions.

When we reinhabit ourselves with appropriate support, we may feel freed from restrictive personal fears or societal norms. As we move expressively and feel our way physically and emotionally through situations and relationships, dominant inhibitions are revealed. Movement is our birthright and connects us globally to other cultures, peoples, and places.

(*facing page*)
St. Kitts Biomedical Research
Foundation Residences
Whitney Sander, architect

Finding Food

Living in Paris my junior year of college, I didn't know how to eat. Having been provided with meals at home and at school, feeding myself was new. At the student cafeteria where we were given passes to eat cheaply, my first meal was *cerveaux* (brains). I never returned. When I tried shopping at the local charcuterie, I mistakenly purchased something with hairs protruding: it was tongue. I settled on street food: delicious cheap baguettes with slices of Camembert cheese slathered on top. I went from 112 pounds to 92 in three months, unaware of the dangers of leanness as my body transformed into a fast-paced, sculpted Parisian "look." Even though I was hungry and constantly on edge, extreme thinness had its euphoric, addictive attributes. Fortunately, I hitchhiked with two friends to Italy over spring break, eating and enjoying! And in June, before returning to the U.S., I visited Germany, where dark bread and beer returned me to sturdy. Every cultural cuisine has its impact; learning how and where to eat is part of arriving.

Reclaiming our whole-body resources and expressivity requires bravery as well as curiosity.

Connections

Today, the invitation is to reorient to ground, space, and breath many times during your day to refresh ease and efficiency in your body. Ask yourself, "Where am I?" And respond, "I'm right here." These words, this meditation, can be helpful in anxious moments by linking movement, breath, and thought for easeful arriving and exchange. Embodied authenticity is the foundation for effective, expressive communication within appropriate social and cultural norms. While sensitive to the values and traditions of others, you remain coherent and resonant within yourself, modeling presence through every cell in your body. ◉

TO DO

FEEDING THE TONIC SYSTEM (CARYN MCHOSE)
10 minutes

Relaxation is preparation for arriving in a new place or challenging situation. That includes refreshing the eyes.

Standing:

- Bring awareness to your feet, palms of your hands, and crown of your head.
- Add *peripheral gaze*: Look straight ahead, relax the back of your neck, and then spread your vision to the furthest edges of your visual field (without moving your eyes or turning your head). Peripheral gaze opens your perceptual lens to the broadest view and the oldest parts of your nervous system. It feeds tonic system sensory cues, answering the foundational question "Where am I?"

Lying on your back with your feet against a wall:

- Feel and relax into the full support of the ground. Let the soles of your feet actively sense the wall.
- Keeping the tonic system nourished, visualize (without moving) that you are going to lift one hand to reach for a glass of water. Note how the state of your body responds to just the idea of a movement—the pre-movement or body-based preparation. Then reach with your hand and arm, maintaining connection to ground as you reach into space.
- Repeat and invite relaxation as pre-movement.
- Now, visualize or imagine that you are going to speak your name to a group. Notice any increased tension in your body, pause, and start again. Then speak, maintaining orientation to ground and space while encouraging relaxation as pre-movement.

- Visualize that you are going to meet someone for a challenging conversation. Relax your jaw as you push the "pause" button between intention and action, so you can notice the pre-movement. Then speak.

In a group: Walk through the space, then pause in front of a partner and orient to ground and space in the presence of one another. Speak your name (introduce yourself), noticing pre-movement. Continue on to another partner.

TO WRITE
I NOTICED, I DIDN'T NOTICE
5 minutes

> *Arriving in a new location is complex.* Describe a specific place you have traveled, starting with: "When I arrived, I noticed . . ." (3 minutes). Repeat this phrase like an incantation, starting each line with this phrase. Then change to: "When I arrived, I didn't notice . . ." (3 minutes). Read aloud and listen to your stories. So much is happening when we arrive in a new place, and we can only take in so much at a time. As we stay "tuned in" to a location over different days, more layers become apparent.

Transitions
Many personal conversations or professional consultations involve intimacy, speaking directly without too much filter. Creating transitions between arriving, entering the exchange, and departing, requires defining beginnings and endings. Within specified time limitations, you have to both establish and end rapport. My colleague, a somatic (body-centered) therapist, describes skills for creating clear boundaries with clients. "Breaking rapport without losing trust—you can change your eye focus, turn away, or create spatial distance. The important thing is to be clear; this is a transition."

Day 7

Anatomy is the language that everybody speaks.— Tom Myers, *from* Anatomy Trains

Power and Perspective

Professional alignment can involve institutional frameworks or "divisions." Because my book *Body and Earth* required an overview of various scientific disciplines, I went door-to-door in the imposing science building consulting with faculty in physics, geology, zoology, biology, psychology, and neuroscience to understand their specific terminology and perspectives. They didn't agree with each other, and hierarchies existed. For example, physics as the "oldest" science assumes its role as the basis for all the others, and the field of psychology can be considered a more recent, culturally variable addition. Yet essentially, we are talking about the same planet, the same species of humans, the same concerns, from different perspectives.

Alignment

Alignment is relationship, not position. It is how we carry ourselves in the world for efficiency and ease, as a communication strategy, and for fun! Sometimes our postural habits are developed from familial or cultural patterns, from life situations or training techniques, or influenced by health or structural uniqueness. Regardless of our postural history, alignment involves balancing the weighty vital contents of our skull, ribs, and pelvis over the full length of our feet—or the base of a chair—in relationship to gravity. It also involves inner alignment, finding coherence between brain waves, heart rhythms, and breath.

Alignment reflects our stance on life. We cue communication by our posture. We can tell from a distance whether someone is confident or insecure, happy or challenged, open to conversation or enjoying inward reflection, and all the subtle and not so subtle shades in between. Multiple silent-level nonverbal cues feed into body image, including alignment, facial expression, breath, eye contact, gesture, movement, and touch—along with relationships to space, timing, and physical adornments.[1] We can also misread nonverbal cues; intercultural meanings vary, while biological imperatives for expression are consistent. Coming into alignment with each other is a whole-person process that requires embracing the foundational continuum between commonality and individuality that underlies communication goals.

Alignment also involves agility. In many communication scenarios, we have to turn, dodge, reach, bend or otherwise adapt metaphorically or physically to interactions. If we are well aligned and resilient, sudden shifts are responded to with fluidity. The bodily core remains spacious and the vital organs free from compression. We also have less harmful impact on the people around us and the places we inhabit if we channel our energies effectively. Watch animals in the wild to notice how invisible their impact on surroundings can be; they can respond to life-sustaining needs with fierceness and fast maneuvers as well as nurturing behaviors.

FACTS

There's a dialogue between *structure* and *posture*. Our bones, muscles, and soft tissues structurally affect alignment—allowing weight to pass through our bodies to the ground. The way we inhabit this structure is our posture— how we present ourselves moment to moment in our lives. Posture affects structure and structure affects posture, and both can change.[2] For example, if you have bowed legs, scoliosis of the spine, or move with the aid of an external device, your posture is affected. Likewise, if you constantly sit slumped at your desk or staring at a screen with minimal movement, structural components like bone density and muscle tone deteriorate. However,

Bandaloup in
Bound(less) (2011)
Artistic director
Amelia Rudolph,
Sleepless Night
Festival, New
World Center,
Miami, Florida
Photograph
© Atossa Soltani

Forward curve of lower spine
Photograph © Angela Jane Evancie

when you begin a movement practice, the body reinforces structural components and builds tone to support what you are asking it to do; if you need more strength and vitality, it is available.

Although verticality is not our only goal, it is one essential dimension of our bipedal stance. Within efficient vertical alignment, the skull, thorax, and pelvis are balanced around an imaginary plumb line or vertical axis. If we draw three ovals representing these three body weights and connect them with a vertical plumb line from the top of the skull to the feet, we have a basic diagram of dynamic standing alignment. From a side view, the front of the body is also called the *anterior surface*, and the back is the *posterior surface*. The three body weights connect with a series of reversing curves: anterior at the neck, posterior at the ribs, anterior at the lower back, posterior at the pelvis (sacrum), and anterior at the small curve of the coccyx—the ancestral tail. Our spine is built for movement. The pelvic floor creates a webbing linking the forward-curving coccyx to the front surface of the body and the abdominal muscles.

Weight passes down the sturdy bodies of the twenty-four spinal vertebrae, in front of the spinal cord in its fluid-filled cavity and spreads through the sacrum. Ideally, standing postural alignment allows the nervous system connection to the brain to be free, like a horse's tail, while weight support is transferred down the front of the spine through the vertebral bodies and discs toward the ground. This occurs in conjunction with a span of connective tissues offering spatial tethering throughout the fluid body.[3] For efficiency in most bodies, the forward spinal curves (neck and lumbar spine) touch but do not pass the invisible plumb line in front. Our heads are heavy, weighing thirteen to twenty pounds. If the head is habitually held forward, thrusting tension into the upper back and neck, the pelvis will counter-balance as a structural compensation.

Consistently standing with the pelvis thrust forward results in short, tight hamstrings and overstretched hip flexors. Muscles and tendons are forced to do the work the bones should be accomplishing, and energy is wasted. You have to come back to center to begin moving, which is inefficient. Instead, it helps to imagine the head suspended by a string from the ceiling or sky, elongating the spine and allowing the head to float upward, extending all the parts toward balanced verticality without rigidity and while remaining connected to ground.

There is no one right position for alignment; every unique body has adaptations and personal patterns may be essential. The goal is simply to use your boney scaffolding to transfer weight to the ground in a way that is efficient for you, while allowing the more elastic tissues of muscle and fascia to elongate spatially, so the organs are free to function. The image of a tent might be helpful: the central pole supports, while the tent ropes are attached to stakes in the ground for tensile strength and support.

Our feet are the foundation. When standing on two feet and lifting an arm, the first muscles to fire (contract) are the soleus muscles in the back of the lower legs, which connect us to the ground for stability. This gives the other muscles something to pull against. There are twenty-six bones in each foot that allow for both stability and mobility, plus shock absorption. The spaces between the bones are loaded with sensory receptors that inform the brain of even the subtlest dynamic shifts between body and earth. As we relax our feet, shins, knees, thighs, and belly, we move beyond tension, freeing the energy in the body that is bound up in tight muscles. Postural tone in the tissues indicates our readiness to engage as appropriate for the situation at hand.

When our bones are balanced in efficient alignment, we feel less. In fact, we may feel as though we are not doing enough. People move to feel themselves working. But no sensation is the sensation of ease. If the bones are aligned, all the other systems can function more effectively.[4] Since the human spine is a series of balanced curves (it's not straight!) swaying over the base of our feet, mobility supports stability. (If we are pushed when rigid, we'll likely fall. If we are responsive, we can adjust.) Humans are built for movement. There are no entirely straight bones or flat surfaces in the body. Responsiveness is the ability to respond. It gives us agency over our choices, allowing our amazing human mobility its full dimensionality.

Our bony skeleton partners the multidimensional complexity of our nervous system. Imagine *our* complex nervous system in the four-footed structure of a horse. Movement options would be too limited. Humans can twist and spiral to move in any direction with ease, and thoughts take

Tonal Readiness
Chris Aiken, teaching Scientific Foundations of Dance at Smith College, discusses body tone in alignment as the ability to respond appropriately in any situation: "Without sound principles for working on ourselves, it is very easy to be counterproductive, layering one level of tension on top of another." Alignment involves being conscious of "a sense of weight and its relationship to how someone perceives space and support. It is challenging to be resilient in your weight if you are overly oriented up or down. And often the sensation of weight is being smothered by an excess of tension."

Naming

In New Mexico, the shop owner selling us chilies assures us he is not Mexican, he is Catalonian—from Spanish heritage. At the Museum of International Folk Art in Santa Fe, we view an exhibit of the history of flamenco from Spain to New Mexico. It details the flight of both gypsies and Jews from Spain during the Inquisition; those who fled to Mexico were sometimes tracked down by the Inquisitors and tortured or killed, so they had to keep going north, founding New Mexico. Naming was used as a disguise of origins and nationality, not identification. Who you aligned yourself with in this context had impact.

us into the past, present, or future. When we restrict our inherent capacity and requirement for multidimensional agility by sitting too long, inhibiting expressivity, or wearing constricting clothing, our nervous system can be compromised—leaving us feeling overwhelmed, depressed, or indecisive. Get up and move, stretch, and dance around, and you may just feel refreshed.

Although we often think of posture as muscles and bones, all the cells and body systems are involved and impacted. Most importantly, the vital organs need space to move, process, and respond. Organs have weight, take up space, and have motility; they rotate, expand, and condense. Open two bodies in surgery, and the organs may be in quite different places and relationships. The gut tube in particular affects alignment and postural tone. Extending from the mouth all the way to the anus and pelvic floor, it takes in nutrients, absorbs and releases what is not needed. A spacious core allows essential strength and vitality in each of our fundamental organs: heart, lungs, liver, stomach, spleen, and all aspects of our digestive and reproductive systems. In turn, the organs affect alignment through their weighted, spacious presence.

ALIGNING WITH OTHERS

We influence how we feel by how we stand. Skills for effective embodied communication are practiced in many work situations, from business and medical training programs to lively TED Talks. If you assume a particular stance, you will begin to feel the qualities of that stance, stimulating your internal body-specific response. Stand tall, and you feel more confident. This could also be described as "fake it till you make it." Movement and mind are intertwined. Authenticity comes from playing consciously with the range of expression that matches your true feelings. Inner radiance is a result of effective alignment that lets energies flow freely in your body and between yourself and others. How we inhabit ourselves is a choice, and alignment is a foundational resource that supports us in responding to different situations appropriately rather than clinging to fixed behavior to affirm identity. We choose how we present ourselves to others.

Alignment varies in different cultural contexts. You can engage alignment in any relationship to gravity. Try squatting; dancing with your torso forward, drawing your heart toward the earth; or sitting cross-legged on the ground with postural alignment in mind. There are many cultural models for sustaining a balanced, healthful relationship to the earth. You can investigate what different movement forms do with the body: moving low to the ground in Capoeira in Brazil, sitting around a campfire drinking mate in Patagonia, dancing heart-to-earth in Nigeria, or synchronizing hands and feet in a Balinese dance. You can also note when it is culturally appropriate to adapt or adopt a stance, remaining responsive to outer cues.

Patterns that are familiar often feel "right." But they might also be hurting you or limiting your expressive range. There is nothing mild about postural change; alignment can be emotionally charged. You may have been criticized for your posture in the past, or you may have adapted a stance

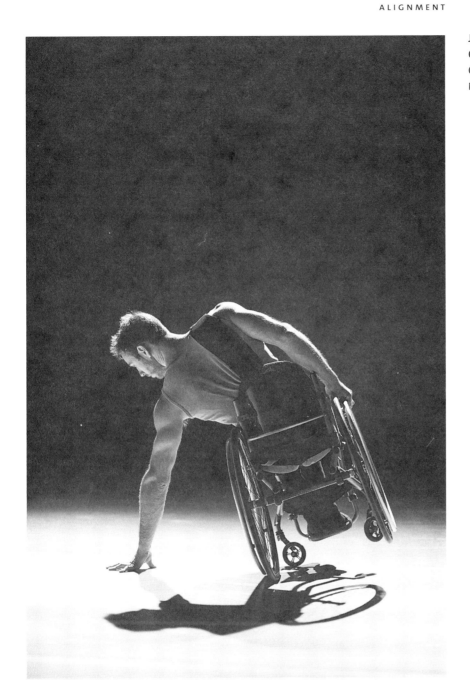

Joel Brown in *Beheld* (2015)
Candoco Dance Company
Choreography by Alexander Whitley
Photograph © Hugo Glendinning

as a strategy for survival (in cases of abuse, lack of safety, or political strife or simply with professional colleagues). You may feel resistance—physical and mental. Sometimes you need outer support, a teacher or coach, to encourage and remind you about effective alignment. But motivation is essentially internal; awareness begins with you.

In embodied communication, the topic of alignment is particularly relevant to inclusivity. Who do we align with based on comfort levels, and how does that affect outcomes? Adjusting our view to include all body types and structural variations based on physical abilities, age, and circumstance, we are reminded that every body is unique. Considering posture as relationship rather than position helps us reflect on how we present ourselves

to others, how we change in different contexts, and why we make those
choices. As we become resilient in our stance on life, we are more able to
communicate with others and respond appropriately and empathetically
to their cues through interpersonal synchrony.[5] We can find ways to go
beyond social and cultural lenses to recognize our basic humanity and
have resources to connect through whole-body communication.

Connections

Today is a good day to enjoy your posture. Begin by noticing your body,
right now, without judging or changing anything. How does this alignment
serve the moment at hand? Then, just for fun, try different postures during
your day and notice their resonance with your intentions. When walking,
try bringing your chest forward; then, try walking with your hips forward—
it feels different. Bring your head over your heart and hips, inviting length
and enjoying verticality. Bend forward, feeling your heartful relationship
with the ground. Arch backwards and open upward to the sky. Just play
throughout your day. Effective posture is related to the moment at hand,
in any orientation to gravity, not a rigid position. 👁

TO DO

THREE BODY WEIGHTS

5 minutes

> *Alignment affects postural tone in the muscles and organs—our
> readiness to engage. In this alignment exploration you may need to
> stand sideways to a mirror to compare what feels right with what is
> visually happening.*

Sitting or standing:

- Establish your base: feel your *calcaneus bones*, the heel of each
  foot. Rock forward over your toes and back to your heels, and
  then center your weight. Relax your knees slightly so they aren't
  pushing forward or back.
- Using your fingers, touch the center sides of your pelvis. Bring
  your pelvis over the center of your feet.
- Touch the center sides of your ribs with your fingers. Bring your
  ribs over your pelvis and your feet.
- Touch the center of each ear (the small flap that covers the hole).
  Tuck your chin slightly and bring your skull over the center of
  your feet, pelvis, and ribs.
- Imagine extending the top of your head about thirteen inches
  above your skull and draw a small circle with this point on the
  ceiling, like it is the top of a pen or pointed hat.
- Touch the top of your head and imagine a plumb line dropping
  down through these primary body centers (skull, ribs, pelvis),
  through your feet, and down into the center of the Earth (where
  all our human plumb lines would meet).

- Then practice balancing the body weights: shift your weight into your familiar "hang out" posture, then realign. Start with your base and build up sequentially through the body weights.
- Practice balancing mobility with stability, refreshing your alignment in relation to people and place.

The spine is a series of partnered curves; it is not straight. Aligning through the bones of your skeleton in relation to the soft tissues of your body offers optimal support.

TO WRITE

MY STANCE (WHO DO I ALIGN WITH AND WHY?)
10 minutes

We've all heard comments about our alignment. Friends recognize us from the ways we stand or sit. Write a few stories about how other people view you and how you perceive yourself. Consider alignment in relation to the rest of your family: do you share postural habits? Who has the lead energy in your family or current work situation, and how is this reflected in your posture? Who do you align with and why?

Day 8

There is a sonic lineage to which we all belong. — Meklit Hadero, from the TED Talk "The Unexpected Beauty of Everyday Sounds"

Breathing

The giant weeping willow tree on the farm was my dollhouse: I would shimmy up the wide trunk and set up my imaginary world on the broad branches. The veil of drooping limbs gave the impression of enclosure. When I fell out, *whomp*, I couldn't breathe. The "wind" was knocked right out of me. As my sister stood staring, I remember vividly the life-and-death pause of no air: would I ever breathe again? Like a later experience of near-drowning, one way of learning about breathing is by not. I don't recommend it.

Breath and Voice

Breath is exchange. We share it with others, including our plant and animal relatives. It is our first relationship with life and our last. It is part of sobbing, laughing hysterically, and silence, depending on what is happening in our lives. Breath is the basis of communication, both verbal and nonverbal. It also informs us about the quality of our environment, including air pollution, weather patterns, and the impacts of people and place on our health. It is a link to spiritual and mystical dimensions, amplifying the ephemerality and mystery of life.

We often take breath for granted, yet we can't live without it. It is the hum and purr of life happening deep within us, essential for every cell and tissue. If you don't have enough breath, you can't think effectively, and you can't speak or find your voice. Breath is a guide and tether to awareness and consciousness (it's always happening), and it is the first link to experiencing energy in the body (you feel it). It occurs both consciously (you can pause) and unconsciously (without your awareness). Ultimately, breathing is beyond our control; if we don't breathe, we pass out: the body takes over and breathes us. But it can also be coached into optimal efficiency.

In communication, we can notice the primacy of breath. Our breath rhythms cue the degree of ease or discomfort we feel when interacting with another person. Breath made audible through vocalization informs others about the tone of our conversation—do we mean it or not; are we threatening or friendly? Ultimately, breathing patterns communicate our degree of engagement: are we supportive of what we are saying and doing, or conflicted? Possibly no other aspect of embodiment can change our physical, mental, and emotional effectiveness in communication more than breath.

FACTS

Breathing habits are established at birth and can change throughout our lives. Patterns can reflect the rhythm of the mother's breath in pregnancy, the birth process itself, or impressionable events from later in life that become fixed in the neural circuitry. The *breathing rhythmicity center* in the brainstem (top of the spinal cord) monitors this basic pattern. Work situations and cultural norms impact breathing patterns, along with our levels of comfort, need for control, and opportunities for expression.

What is involved? Oxygen essential for human life is produced by plants, which utilize the carbon dioxide that we exhale. With each inhale, oxygen-filled air becomes part of our internal body; with each exhale, carbon dioxide-filled air returns to the external environment, in constant exchange with the places we inhabit. If air only gets to the lungs in this process, you die. The heart has to circulate and spiral the oxygenated blood

Meklit Hadero
Photograph by Marc Hors Domingo

throughout the body so that every cell "breathes" in a process of fluid exchange called cellular respiration.[1]

The primary muscle that contracts to initiate breathing is the musculotendinous *diaphragm*. This double-domed structure within the responsive ribs is anchored by the crus tendons, which attach along the front of the lower spine (lumbar vertebrae) and connect through fascia all the way to the tailbone. In contraction (the *in-breath*), the diaphragm spreads and descends downward toward the pelvis. On the release (the *out-breath*), the diaphragm returns to its double-dome-like shape (around nipple height in relaxation), expelling air from the lungs. The diaphragm, as the floor for the heart and lungs and the ceiling for the stomach and liver, assists

Words and Meaning

After teaching at an international institute, I never assume that English is everyone's first language. Often, I ask participants in a movement workshop to translate a term or phrase that I am using, speaking it in their home language to broaden perspective. "Walking," for example, feels different in Spanish, Japanese, Portuguese, and Russian. Language shapes embodiment. When explorations are only taught in English, students returning to their home countries have not practiced somatic articulation in ways that are useful to the delicious richness of their home tonalities. Movement evocations are subtle; we aren't using words to tell someone what to do, we are inviting and transmitting an experiential process.

in propelling fluids throughout the body.[2] Thus, the rhythmic contraction and release of the diaphragm massages all the vital organs.

Any restriction in the ribs and waist, such as tight clothing or over-rigidity in the intercostal and abdominal muscles, interrupts efficient breathing. As every cell receives oxygen and releases waste materials, the whole body breathes—expanding and condensing with each breath. Notice if you can feel this process in your own body. Remember that the tailbones and pelvic floor are involved in deep, full breathing. (The xiphoid process, which is the "tail" of your breastbone/sternum at the front of the ribs, is also moving.)[3] The mental image you hold of the breathing process can assist or interrupt efficiency. The heart circulates the oxygen rich blood to every cell in your body. If you limit your awareness of breathing to the lungs, you miss essential dimensions.

Some people are "reverse breathers," lifting the shoulders up and flattening the belly to breathe in, so that the diaphragm barely moves. Some people "over-breathe," working too hard to control the process on both the in-breath and out-breath. And some of us are "shallow breathers," barely moving the ribs and spine. When we don't want to absorb the air around us, release control, or be seen, we hold our breath. Although there is no right way to breathe in all situations, it is useful to develop an attitude of open exchange with the environment you are in. Tension restricts movement and masks sensation, numbing us from noticing the sensitive interplay of breath and context.

Breathing affects all the body systems. *Bones* create all of your red and white blood cells in their fluid marrow, including the hemoglobin proteins that carry oxygen. These red blood cells also remove carbon dioxide from the body, carrying it to the lungs to be exhaled. The health of bones affects the health of blood and basic vitality levels. Bones are important for longevity and healing; through visualization we can direct our breath into the bone marrow to encourage healthy blood.[4]

The *brain* uses tremendous amounts of energy, depleting oxygen and glucose levels needed for other organs and the bones. After an extended period of strong mental focus, we may feel tired or fatigued, and the vital organs suffer. Through visualization and relaxation techniques, we can "breathe into" the organs, unwinding tension and fueling oxygenated blood flow in our deep core. Rather than emphasizing the thinking mind, we bring our attention to the belly-mind through breath.

The muscular spirilic heart has to circulate oxygen-rich blood through over seventy thousand miles of vessels to every cell and tissue. There's a link between breath and heart health. Tension anywhere in the body increases the workload of the heart. Most heart clinics begin health initiatives and rehabilitation treatment with deep-breathing techniques to lower blood pressure and take workload off the heart.

What interrupts an efficient breathing process? Fear, restrictive clothing, lack of grounding, emotional uncertainty. Inefficient alignment also plays a role. Muscles, to contract, need something firm to pull against. Most are attached to bones, and bones lever into the ground through our

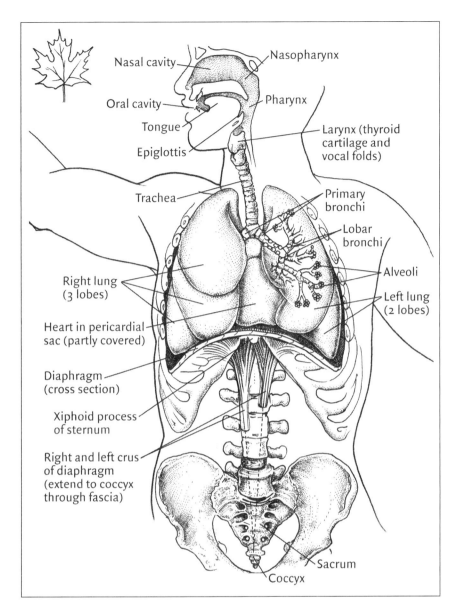

Nasal cavity
Nasopharynx
Oral cavity
Pharynx
Tongue
Epiglottis
Larynx (thyroid cartilage and vocal folds)
Trachea
Primary bronchi
Lobar bronchi
Right lung (3 lobes)
Alveoli
Left lung (2 lobes)
Heart in pericardial sac (partly covered)
Diaphragm (cross section)
Xiphoid process of sternum
Right and left crus of diaphragm (extend to coccyx through fascia)
Sacrum
Coccyx

Organs of respiration
Anatomical illustration by Nancy Haver (2021)

effective orientation. We need grounding to stabilize the bones, allowing tissues to pull against our mobile architecture, not through rigidity but firmness. It is like two trees holding a hammock: there is both rootedness and responsiveness.

It is important to know our breathing habits and to enjoy the pleasure of inner movement. We can use language that is enlivening rather than mechanistic. For example: imagine your ribs as a responsive basket rather than as a cage. There is a glow of inner radiance in an integrated, well-functioning, and healthy system, from the descent and rise of the diaphragm massaging the heart, to cellular responsiveness feeding every tissue. As we place our hands on our ribs (top, bottom, back), there is a spongy, rhythmic filling and emptying of the lungs. Breath is always changing, responding to context and conditions. Without judging our breathing patterns, we can notice and encourage efficiency.

Language and Recovery
A Spanish professor at my college had been a combat soldier in Vietnam. When I asked him why he studied Spanish, he said, "When I got back from war, I didn't want to think in English. I needed a whole new vocabulary and way of thinking. Language immersion helped."

Many vets returning from combat are working their way back into daily life. Voice is impacted and impactful: it is often silenced or comes out as a command. In group dynamics, both can feel oppressive, aggressive. Modulating tone involves calibrating the space between speaker and listener(s). The practice I offer to my class: "Put one hand in front of your face and make sounds that just touch your lips. Then place what you are saying into the space between your lips and hand. Now, touch your hand with your sound or words. Direct sound into the middle of the room, and across the room to an imaginary person—touching their body lightly with your voice, or penetrating, pinpointing." Finally, we expand vocal range to include the whole group, whole room before bringing sound back to close range, to self. It is fun.

Words can hurt; they can also heal. We can choose where we place our voice as well as what we say. One student was particularly relieved; from years in the military, he had forgotten that he has choice.

Everyday Sounds

Meklit Hadero, an Ethiopian-American singer and songwriter living in San Francisco, brings her band to our school to perform. Her TED Talk "The Unexpected Beauty of Everyday Sounds" has inspired many to listen more carefully, while the beats of her music get your body moving. The auditorium fills to the brim for this Intercultural Arts and Social Change event; excited audience members come from all departments as well as the community. "Divisions" merge; music brings us together, especially when masterfully offered.

VOICE

Voice is breath made audible: it is part of our human heritage and sensual terrain. The diaphragm determines volume, the vocal folds create the vibrations, cavities influence resonance, and the mind determines intention and direction. As sound vibrates the inner tissues of the body, we feel what we are saying, touched by the felt sensations that we interpret as emotion. Sound resonates outer dimensions as well, including other people. We touch each other with our voices. Voice describes and articulates the process of exchange. Presence, quality, and tone of voice impact power dynamics: who gets to speak and why.

Humans are hardwired for speech, provided we breathe enough to allow sound to emerge. We map our voice from internal sensations: the way our body feels as it vibrates, resonates in relationship to particular people and places, and shapes who and how we are in the moment. Feeling what we say, cycling between impression (taking in) and expression (giving out), is a fundamentally circular rhythm of communication. If it is all outward, you deplete your resources (leave no space for the other, for exchange). If it is all inward, it is hard to connect (have no space for the other, for exchange). Sound is tactile: you are touched and touch through vibration as you speak.

For sound production, the only muscles necessary (beyond those of breathing) are the tiny ones affecting the vocal cords. Tightening the superficial muscles of neck and shoulders, belly and back, only reduces resonance. As air moves from the lungs on the out-breath, it passes through the *trachea* (along the front of the neck) to the *larynx*. This houses the *vocal folds*—mucosa-lined ligaments or cords that run parallel to each other from the front to back of the larynx, forming a horizontal vocal diaphragm (voice box) in the neck. If these folds are apart and relaxed, the passageway is open, and the air flows freely through from lungs to nose and mouth and is released into the environment.[5]

If, however, there is intent to make sound or speak, the folds are pulled taut, creating a thin space for the air to pass through, vibrating the cords. (If you have a cold, excess mucus irritates the cords, thickening them so that they vibrate more slowly, creating a lower voice.) These vibrations are amplified in resonating chambers of the body, shaped by the soft tissues of the *pharynx* (throat) and mouth, and articulated by the lips, teeth, and tongue to produce audible sound. In speech, vowels vibrate, while consonants interrupt, clip, and shape the flow of the air. Languages have different uses of vocal mechanisms for distinct sounds and flavors.

Vocalizing without language registers in the limbic brain, our "older" emotional, relational brain that evolved with mammals around two hundred million years ago. Simple vocalizations can have emotional cues: warning, invitation, territorial claim. For example, a mother recognizes her baby's cry amid a crowd; yells at football games stimulate adrenalized response; humming in a lullaby soothes. For some of us, freeing vocal sounds can trigger our own protective, defensive behavior in relation to emotional expressivity. We may become tense when hearing any escalation of vocalities. How we interpret sound is based on past history. If yelling was standard communication in your house, it might feel like love; to others it could be threatening. Our relationships to vocal cues play a role in how receptive we are to others, how we speak, how we hear, and how we respond.

Many of us identified our voice around puberty, when our bodies achieved their adult dimensions, and now feel comfortable with the voice associated with those particular sensations. Yet we have changed a lot since then. Rehabilitating our relationship to voice is a process. Simple vocal skills are easy to practice and can have a significant impact on communication and self-identity. Do you recognize your voice at this age? Would you like to increase resonance and adaptability? What personal and cultural contexts have affected your vocal habits? Where do you project your voice? Do you hear what you are saying? Are you sending your voice between you and another person, at their skin, or penetrating and potentially violating their internal experience (taunting or bullying)? Like physical touch, vocal touch has implications and is affected by intention. Choice is involved.

Honoring one's own internal "shutdown" of voice, the hesitation or deep resistance, is essential. It's important to recognize that people around the globe since the origins of our species have been tortured, exiled, and killed for speaking their truth. They are present in our myths, stories, and history books. Joan of Arc and Little Red Riding Hood represent warnings: speak and you may be killed or led astray. Martin Luther King, Nelson Mandela, and John F. Kennedy were models of the risk of big thinking and speaking out. Contemporary women in the news—a woman senator shot, a Muslim girl drenched in hot oil for refusing an arranged marriage—remind us all that there's risk involved in speaking. Our bodies want to keep us alive. When we feel the grip, the strangle, the heat rising, the tension building, the impulse to flee or to silence ourselves, it is well-founded.

Vocal power can be seen as the ability to inhabit voice with authenticity. The whole body is involved in resonance and amplification in relation to the specific context of place and people. This includes allowing emotions, dynamic range, and emphasis to inform expression for clarity and impact. And in speaking with a partner, pauses for breath let you and others feel what you say. This silence is not from repression but is a choice that encourages absorption. Silence is part of speech—it's the space between sound and no sound. For some, it means love, listening, opening to larger dimensions.

Once we identify *where* we are through orientation and alignment, we can engage more complexities: *who* we are, *what* we have to say, and *why*. In this process, we continue teasing biology from biography. All humans on the planet share ancient survival reflexes, developmental patterns, and emotional responses that govern movement and voice in the early months of life. This heritage illuminates our common ancestry—on this planet and/or from stardust. In vocal exchange, we can remember that we are standing, literally, on the same ground. And somewhere, on the other side of the Earth, someone else's feet are matching yours—body to body, supported by planet Earth.

Bad Advice
My neighbor in Florida states that his one claim to fame as a teacher was telling a high school student that she could never make a living as a musician. He encouraged her to go to college for a different profession. Then he gives the punch line: the student was famous singer Diana Ross.

Connections

Walking through doorways today offers opportunities to refresh breath and voice. What might you want to leave behind from the past? How do you feel in the moment of transition? What are you arriving into or moving toward? Stepping through each opening is an opportunity to take a deep full breath, clearing old energy and making space for the new. It is also a good time to enjoy cellular breathing, pausing to feel resonance within your whole embodied self. 👁

TO DO

AUDIBLE SIGH

5 minutes

> *Grounding before sounding supports easeful vocal work. Imagine you are in your own personal "safe place" where you can make sound without any judgment or restriction.*

Seated or lying, eyes closed:

- Feel the ground supporting you as you bring your awareness to your breath.
- Take a full, deep in-breath, and on the out-breath allow a relaxing audible sigh to fall out with your breath. Repeat three times, dropping the image of safety into your body on the in-breath, sighing on the out-breath.
- Allow your vocal folds to vibrate without adding any tension to muscles in the rest of your body. If you tighten in your pre-movement, pause and reorient to ground and space, giving more support to easeful vocalization.

Sound Play

- Now, play with sounding. Place your hands on the bottom of your ribs and trace their circumference. Feel this attachment site for your *breathing diaphragm*, which creates a horizontal trampoline through the body. Encourage the spinal movement of breathing all the way down to the front of your tailbone (coccyx).
- Make the sound "ha" with your breath. Repeat "ha" several times, as if you are bouncing the sound on your breathing diaphragm. Find a speed that is comfortable to you.
- Move the sound anywhere in your body—bounce sound into your pelvis, your feet, your skull.
- Change the position of your face muscles with each out-breath, allowing the sound to alter with the shape of your mouth.

- Add the body, inviting any movement that feels good, enjoying audible sound to move you without restriction. How does movement affect your sound?
- Pause and notice what you have stirred up with sound. Let yourself be nourished by the lingering vibrations of sound.

TO DRAW AND WRITE
MY VOICE (ALEX DRAPER)
45 minutes

Your voice has a history. Give yourself time to explore. Get two pieces of paper and lots of colors of pens/markers/crayons. Draw your voice as it was in the past (5 minutes); turn the page over and draw your voice as you'd like it to be (5 minutes). Then write your history with voice. Begin with childhood and write any memories and experiences related to voice up to the present moment (30 minutes). Read aloud, and sense and feel your voice (5 minutes).

Fabiano Culora
London, England (2015)
Body and Earth: Seven Web-Based Somatic Excursions
Still image from videography
by Scotty Hardwig

Balancing the Nervous System

The nervous system governs our actions and reactions. From conscious planning to life-saving reflexes, our multidimensional nervous system partners our ability to make choices between action and reaction, fragmentation and coherence, freshly informed decisions and habits in communication. To effect change in personal patterns and group interactions, we can also consider the nervous system as a "friend" and enjoy constant conversation with its various layers and parts. The invitation is to stay in rapport with our multiplicity in the context of other people and places.

Negotiating a relationship with the nervous system can involve re-appreciating all the parts. This includes the evolutionary heritage that underlies our current structures and behaviors and the language we employ to describe interactions. Rather than using terminology to distance us from our bodies, we can use naming to encourage familiarity. For example, differentiating the three components of the autonomic nervous system can enhance a sense of humor about the multiple internal "voices" we listen to (or ignore) that govern our lives. As we choose to balance different needs and desires, we can notice what parts cue sleep and digestion, activity and expression, and complex thought processing and imagination to support their efficiency. Returning to a sense of wholeness by dissolving fragmentation and disorientation is the ultimate goal.

A compromised nervous system and "dis-ease" are common results of our fast-paced lives, limiting efficiency and pleasure in encounters. Unmanaged stress affects the chemical balance of our bodies, as does food, the company we keep, and the choices we make around activation and response. The immune system in particular is impacted as we face sensory and emotional overload, numbness, or unmodulated reactivity. We need a daily practice to clear stress, feed energy (oxygen and glucose) to our full bodies (not just the brain), and open us to our fullest potentials.

It is useful to remember that the nervous system works in tandem with another governing system, the endocrine system, to facilitate and supervise whole-body activities. The nervous system communicates through electrical signals (impulses) generated by nerve cells and transported along nerve fibers; the endocrine system communicates through chemicals (hormones) transported through the body fluids. Through this combined electrochemical signaling—sharing or transferring information through electrical signals that are translated to chemical signals and vice versa—receptors and channels act as mediators between these two modes of communication. One informs the other in a dynamic feedback loop. As an interconnected whole, the neuroendocrine system helps us unveil the resources and habits that can connect us to or distance us from the world and each other.

Sometimes the brain is the last to know. — Bonnie Bainbridge Cohen, from "Exploring the Embodiment of Cellular Consciousness through Movement: A Body-Mind Centering Approach"

(*facing page*)
Paul Zivkovich
Photograph © 2014 Lois Greenfield

Raw

In our young dance company, I would often die. Choreographer Katherine Sanderson would have me lie on the floor, still, while others moved around the stage. Or I was carried limp in the arms of a running dancer. The motionless figure often captures the viewers' attention—what's wrong? This creates a projective field—viewers get to empathize with stillness or collapse without actually going there.

FACTS

The nervous system consists of the brain, spinal cord, and an abundant network of neurons extending to (and from) all regions of the body. Generally divided into two parts for study, the *central nervous system* includes our brain and spinal cord, and the *peripheral nervous system* includes the nerves and ganglia outside of the brain and spinal cord. The peripheral nervous system has two distinct components: the *somatic nervous system*, which affects voluntary movement (our sensory-motor responsiveness), and the *autonomic nervous system*, which affects visceral functions (such as digestion, heart rate and reproductive activity). Considered as an interconnected whole, the nervous system records, orchestrates, interprets, plans, and remembers our movement life.

CENTRAL NERVOUS SYSTEM (CNS)

We can begin this inquiry into self-awareness with the central nervous system, which is encased in the bony spinal vertebrae and skull.

Spinal Cord (reflexive movement)

The *spinal cord*, located within the vertebral canal, is responsible for reflexive movements as well as for transmission of impulses to and from the brain. The fastest, most efficient neuromuscular path in the body is a self-protective *spinal reflex*. You withdraw your hand from a hot stove before the pain or action even registers in the conscious mind. This reflex arc makes an expedient loop from the sensory nerves (in the skin or deeper tissue) to the spinal cord, returning through a motor nerve to signal a muscular response. Simultaneously, neurons within the spinal cord inform the brain about what is occurring. "Ouch!"

When communicating, postural reflexes are the foundation for movement and speech. Reflexes and human developmental patterns orienting the spine in relation to gravity are encoded from our evolutionary heritage and are developed in early childhood; they underlie the demands of an upright stance and the complex interactions of living in social groups. As the fetus develops in the womb, the primary centers of the brain mature and develop neural pathways, connecting various motor centers located in the more ancient parts of the brain to each other: the spinal cord, brainstem, midbrain, and cerebellum. Collectively, these essential areas are often referred to as the "back brain," "hind brain," or "coordinative brain." At birth, full-term infants have multiple reflexes involving the mouth, face, and whole-body that form the basis for survival. These almost-instantaneous, internal coordinative pathways allow us as adults to walk through familiar terrain talking to a friend without staring at our feet. Our bodies adjust unconsciously to uneven surfaces and obstacles along our path through our kinesthetic, body-level intelligence. Rigidity in the spine is a sign of distress or distrust in communication.

Brainstem (central pattern generator)

The *brainstem*, an enlarged portion at the top of the spinal cord, is a central pattern generator for vital organs. It includes the midbrain, pons, and medulla oblongata, and forms an essential link between the spinal cord and the cortex (brain). Among its many roles, the brainstem directs and maintains consistent patterns for breathing, heart rate, and organ function. You can't think or move clearly with insufficient breath; the body prioritizes oxygen over everything.

For efficient communication, we can cultivate consistent, saturated breath even if situations become stressful. When speaking, breathe before delivering the thought. Breath is inspiration. This sets the nervous system at ease to focus on other complexities of presentation and interaction. Shallow or unsettled breathing undermines our ability to focus, restricting higher thought processing.

Midbrain (attentional reflexes in relation to context)

The *midbrain*, a one-inch section of the upper brainstem, regulates reflexes for focusing attention toward outward stimuli: reflexes such as turning the head and whole body in response to a startling sound. Attentional reflex patterns create readiness to respond. They are pre-emotional survival patterns. For example, the Moro reflex in babies initiated by a sudden movement or surprising sound is a precursor to the startle response in adults.[1] Sleep and wakefulness are also regulated in the midbrain.

In communication, the midbrain supports fast, whole-body shifts of attention. Moving beyond self-involvement, it keeps us located in relation to context. The whole self can instantly focus: eyes and body turn toward the stimulus, and the palms of hands and soles of feet activate, preparing for a quick response. Focusing whole-body attention in relation to outer cues is spontaneous, unselfconscious, and direct. There's no ambivalence or hesitation. Restricted or indiscriminate attentional focus limits engaged interaction and undermines completion of projects and conversations.

Limbic System (processing emotions)

The *limbic system* surrounding the base of the cortex is largely responsible for processing emotions, affecting memory and learning. Often referred to as the emotional brain, the limbic ring is not one structure but a collection of structures present in animals, and it is particularly developed in mammals with the relational requirements of live births, long parenting cycles, and social group living arrangements.

Important in communication, emotional-relational connections bring feelings into conversation: we care about what's going on. The limbic brain is the story part of the brain: we want to know what happens next. Overall, the limbic system is involved in our emotional intelligence, which includes alertness, motivation, passion, and affection. Challenges with emotional-relational modulation affect communication. Some situations require relational interactivity for building social capital; others need careful and

Practicing Empathy

Doctors-in-training have a short course on empathy. I have been told there are five primary steps: 1) make direct eye contact; 2) greet and say the patient's name, and introduce yourself and your role (Hello ____, my name is ____ and I'm your doctor); 3) repeat their name three times during the conversation; 4) touch them on the arm or somewhere on their body in a safe way, or gesture to break down the space between you; and 5) end by asking if there are any questions. When I go to my dermatologist— who has been rated "best in the Valley" for several years in a row by patients—I watch this sequence in action. I like her!

Something New

I make a birthday vow (to myself) to do something new each day, for a year. In the beginning, this is easy. By dinnertime, if I haven't found something new, I choose a new recipe. Today, my challenge involves turning right down the beach instead of left. It's so simple but engages an entirely different community of people. Stretching the container involves valuing the unfamiliar, waking up my "amygdala activation" for "new." Try it; it's harder than you might imagine.

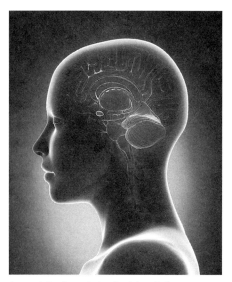

Amygdala: female brain, lateral view
Image: iStock

respectful distance, especially in intercultural contexts where social norms are not yet understood. Versatile communicators want access to emotional discernment—being able to choose the appropriate amount of energy and expressivity for the context—so they can listen as well as share their views.

Amygdala (responds to newness or threat)

The *amygdala*, an almond-shaped structure in each brain half, is considered the "emotional sentinel" of the limbic brain. Often studied in terms of fear, the two amygdalae register anything new or unusual in a person or the environment that might be dangerous.[2] Screening for emotional relevance at the level of survival, the amygdalae determine the speed and complexity of processing, sometimes overriding higher thought processes. When healthy, the amygdalae keep you alert for safety; hyper-vigilance creates stress. Amygdala activation is based on your history. If you've traveled often to other cultures or been engaged in dangerous situations, your "newness" threshold is higher than those with more stable conditions. Repeated stress can cause dysregulation of this system.

As we communicate, the question becomes: How much newness do we need to stimulate alertness and keep others awake to our ideas? And how much makes us, and those around us, retreat from the conversation? If the presentation is too safe, listeners settle back in their chairs and snooze—they've heard it before. However, an overactive, overstimulated amygdala creates hypervigilance. Exploring the edge between comfort and fear lets the amygdalae become a little less reactive, so we can sustain curiosity.

Cerebral Cortex (thinking and processing)

Decision making, planning, noticing, verbalizing, and remembering are all functions a healthy *cortex* can carry out when it isn't overwhelmed with micromanaging all the lower functions of the nervous system (that are spatially below when standing, and evolutionarily earlier in development). Paired cerebral hemispheres govern these processes: each cortex half or lobe has its own responsibilities, but they also provide duplication in case of injury. Neuroplasticity allows some functions to be taken over by other areas when necessary. Overall, the cortex harbors a variety of subdivisions that support different types of higher-level cognitive processes.

In interactions, the right brain is involved with real time and place, and the left brain with abstraction of real time and place.[3] In other words, the right brain (sometimes called the "creative brain") engages the moment-to-moment processes of a situation, and the left brain is involved in composing, translating into language, and talking about the experience. The *corpus callosum* is an arched bridge of nerve fibers that links both cerebral hemispheres with approximately two hundred million nerve fibers facilitating communication. This connective pathway is important for integration of various brain functions and has been shown to be enlarged in exceptionally creative and brilliant individuals, like physicist Albert Einstein.[4]

In communicating, we can make decisions, plan ahead, remember, and be clear and thoughtful in choices, actions, and feedback. Overthinking

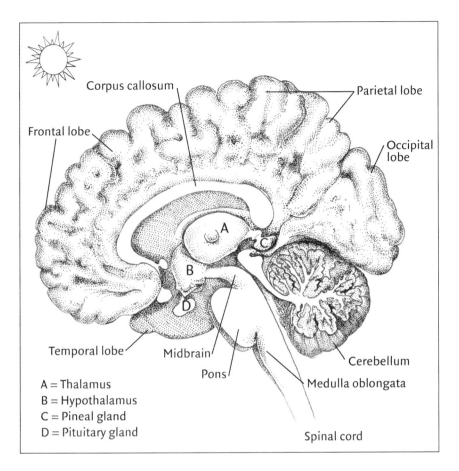

Right cerebral hemisphere
(medial view), from an MRI of a
twenty-four-year-old female dancer
Anatomical illustration by Nancy Haver (2021)

can limit engagement and compromise good listening skills, interrupting rather than supporting whole-body intelligence.

Review of the Central Nervous System

We nurture our brains; where we live and what we do shapes who we are. Exhaustion of the central nervous system is a factor in many contexts. Take a moment to review, embody, and absorb the information so far. Stand, find ground and space for support, and shake out the spine, allowing the central body to be mobile and responsive (spinal reflexes). Take a deep breath and cue relaxation (brainstem). Focus on outer stimuli and refresh perception—be surprised and respond reflexively to the changing environment around you (midbrain). Add emotional connection to what you are seeing and experiencing through memory and association (limbic brain). Then think about the process. Invite innovation (cortex). What else might you add to your exploration? Naming, distinguishing, and embodying these different parts of the central nervous system supports adaptability in complex settings. For example, some situations require more clarity and leadership, others more empathy and receptivity.

AUTONOMIC NERVOUS SYSTEM (ANS)

The *autonomic nervous system* is the medium through which both activation and integration occur; it's our source of vital, visceral responses. One

Demonstrating Understanding
Interpersonal neurobiologist and popular author Dan Siegel demonstrates the central nervous system by holding up his fist. Curling his fingers around his thumb, he explains that the cortex is the outer layer, the limbic brain is the thumb, and the wrist and forearm are the brainstem and spinal cord. It makes me smile to see/feel this inner complexity demonstrated so clearly. Wrapping my fingers around my thumb lets me enjoy the evolutionary layers that are helping me see, feel, and coordinate movement as well as understand my life.[7]

of the signposts of inhabiting ourselves with more ease through a balanced nervous system is that we can be more fully present in interactions, welcoming others into conversation and collaboration.

FACTS

The autonomic nervous system is composed of nerves and ganglia located along the front of the spine, anterior to the vertebral bodies. It also lines the digestive tract. Governing the vital organs and glands, the ANS affects heart rate, breath rate, and digestive and sexual functions; channels of information flow throughout the body and brain. The ANS can function rapidly and continuously without conscious effort, regulating the visceral survival processes that maintain the body.

The ANS has three interconnected components: the *sympathetic, parasympathetic*, and *enteric nervous systems*. The sympathetic division of the ANS stimulates the body toward activation and engagement, the parasympathetic toward cycles of rest, digestion, and integration. The enteric, the most primitive aspect of the nervous system, is the "brain in the gut," the local nervous system of the digestive tract. In effective functioning, the sympathetic, parasympathetic, and enteric nerves work together, not antagonistically, to coordinate and modulate body functioning for optimal vitality and health.

In communication, the ANS is considered the location for "centered" energy. Related to the *hara* in aikido and the *dan tien* in tai chi and qi-gong, autonomic survival responses in the belly-brain in particular support integrated movement, underlying a sense of integrity, coherence, and authenticity.

Sympathetic Nervous System (SNS)

During normal situations of alertness, the *sympathetic division* supports clarity and directness of action. Blood flows toward the skeletal muscles for outer engagement with the environment. But during what is perceived as a stressful or life-threatening situation, the sympathetic nerves trigger what is often called the fight, flight, or freeze (*fff*) responses. The body prepares for activity by reducing the non-urgent functions of digestion and heightens muscle and sensory preparedness. This includes dilation of pupils, decrease in digestion, increase in blood to skeletal muscles, and increase in blood glucose concentration in preparation for activity. Bladder muscles also release to reduce energy expended—there is typically the need to pee and clear the digestive tract before a public presentation or competition.

Contemporary descriptions of this fight, flight, freeze autonomic response model have revised the terminology to include tend-and-befriend. Bonnie Bainbridge Cohen offers an additional view, reminding us that there are other embodied possibilities when the autonomic nervous system registers. You can pause and return to cortical processing to take in more information from present-moment awareness. "In fight, movement is directed spatially in the direction of a perceived oncoming force. In flight,

Friendly

When I teach about the "fight-or-flight, freeze-or-faint, and tend-and-befriend" responses of the autonomic nervous system, I am reminded of this story: I was a junior in college, leaving for a year abroad in Paris. Our group met in Chicago, and we were assigned individual hotel rooms. As I went upstairs, I remember a man in a black suit as one of several passengers on the elevator. When I unlocked my door and entered my room, he put his black, shiny shoe in the door—emotional memory, stored in detail—so that it wouldn't close. Then he forced his way in, locking the door behind him. What surprised me most was my response: I began to talk, launching into a tirade about how it is men like him who make girls like me have to be afraid, and I continued ceaselessly in an irritated and commanding tone. And he left.

In that moment of decision, my nervous system registered that I couldn't fight (he was stronger), couldn't flee (the door was locked), and it was useless to freeze, faint, or befriend, so I talked. This activated a verbal response; our capacity to overtalk, make facial gestures, or otherwise verbally defend ourselves is a survival mechanism that many of us use all day long. Now, I try to notice when I am speaking from my central nervous system in a balanced state, and when my voice and face reflect sympathetic over-activation. Notice for yourself if you consider talking part of your defense system.

movement is directed spatially in the opposite direction of a perceived oncoming force. In freeze, there is no movement through space, either by fully reducing our postural tone and collapsing, or by increasing our postural tone and rigidifying. Ease allows yielding into ourselves, gravity, and space as a foundation to move three-dimensionally through space in infinite possibilities in response to a perceived oncoming force from any direction." By accessing conscious movement choices, you may find more easeful pathways to act rather than react. She describes this process as fight, flight, freeze, and ease.[5]

In communication, the initial edge of sympathetic activation brings alertness, dynamic range, and externally focused presence to interactions. With ganglia located in the thoracic and lumbar regions of the spine, over activation of the sympathetic nervous system can result in hyper-extension (arching) or bracing in the lower back. You can calibrate sympathetic stimulation by finding just the right amount of back-body tone to support your postural and gestural clarity without rigidity while activating front-body support. Inviting optimal alertness without strain involves embodied awareness in the actual moment (ground, space, breath), rather than emphasizing fears or expectations about an encounter. Overstimulation results in stress responses, lack of conviction, diminishment of listening skills, and miscalculation of energy.

Parasympathetic Nervous System (PNS)

Relaxation is the inroad to discovery. The *parasympathetic division* is activated when we are ready to focus internally, relax, and digest with time for integration, like a quiet afternoon after a big meal. In this situation, the heart rate and breathing slow down, the eyes relax and water, and blood flow is directed to the lungs and digestive system. As the digestion activates, increase in secretions sometimes cause familiar stomach grumbles. Yawning stretches the interior lining of the mouth and releases jaw tension. Reduction of blood and glucose levels in the brain and skeletal muscles causes drowsiness. This restful digestive process allows recovery and integration for all the body systems.

In communication, the parasympathetic nervous system provides the basic background tone for healthy nervous system engagement. Also called the craniosacral division of the autonomic nervous system, its nerve ganglia are located within the brain, the cranial base (brainstem), and the sacral portion of the spinal cord. Tonal readiness involves relaxation (without collapse) in all the tissues. A softening of the front body may occur, as well as deepened breathing and a sense of receptivity. Relaxing the back of the neck can release spinal tension and encourage easeful rest and digestion. It is often during parasympathetic activation that seemingly disparate thoughts and feelings come together, and we have highly specific, clarifying insights or images. Optimal functional connectivity allows association and connection between brain regions, especially subcortical-cortical connectivity. A cyclic balance between rest and activity is fundamental to the healthy functioning of internal organs, including the brain.

Connectivity
Dr. Stephen Porges reminds us that the human socialization system can be toned, trained, and patterned by playing music—wind instruments in particular. He's a clarinetist and knows you have to modulate breath rhythm while you are constantly communicating to listeners and other players. Breath by breath, there's connectivity.

Yet many of us know the situation of eating a meal and running off to a meeting or event. In this case, the nervous system stimulates the digestive system for processing food, and the sympathetic nervous system is simultaneously activated to deal with high-level functioning in the world. One physical effect of these conflicting messages is indigestion. Continued imbalance can result in deep parasympathetic exhaustion and situations of chronic disease. To encourage parasympathetic states, we can relax tension in the front surface of the body and encourage a spacious quality (with time for digestion and absorption) in our interactions.

Enteric Nervous System (ENS)

The *enteric nervous system*, or the "brain in the gut," is composed of a primitive nervous system called the neural net within the lining of the digestive tract. Part of the autonomic nervous system, the enteric has one unique characteristic: it can function autonomously, processing sensory information (and motor response) to maintain homeostasis with little or no interaction with the brain and spinal cord. Formed by over one hundred million neurons (more than the spinal cord), the enteric nervous system involves two layers of tissue surrounding the esophagus, stomach, small intestines, and colon. Studies show that it can act independently, learn, remember and feel.[6]

In communication, we can reflect on situations where we followed our instinct or gut feeling. We made a phone call at just the right moment, encountered a friend, or met someone at an unexpected time. Functioning below rational thought, our visceral body picks up on cues (information) that may not register in our conscious minds. This "instinctive" aspect engages another kind of intelligence and knowing (shared by our animal relatives), balancing our picture of ourselves as brainy and smart. Various models have been offered to explain the seemingly mysterious process of unconscious knowing; yet in our daily lives, we notice that when we engage all the dimensions of our nervous-system potential, we feel supported by, rather than in conflict with, our deepest motivations. To support the enteric nervous system, we can keep the front surface of the body open and receptive in communication, allowing the enteric nervous system its "voice." Shutting down the enteric nervous system can result in a lack of instinctual intelligence and more discomfort with basic feeling states.

All parts of the peripheral nervous system (beyond brain and spinal cord) are woven together by an interconnected web of fascia. This connective tissue has its own way of communicating as a fluid semiconductor and support system. It can also get stuck together, tangled, squeezed or otherwise limited by tension or inefficient alignment. Holding restricts sensation and flow. Healthy communication at the intrinsic, embodied level requires spacious interaction of all the parts, enabling the web of interconnectivity. If we can experience this fluidity at an internal level in ourselves, we can amplify connectivity rather than resistance or conflict in communication.

Planning
I am a thinker, a planner. When I sit in meditation or enter Authentic Movement, I can cortically plan my whole life as well as other peoples' involvement in my projects. Finding the gut body was particularly revelatory: it let me realize there is more than one way to organize a life.

Embodied intelligence acknowledges and appreciates all aspects of the nervous system and their needs for both expression and recovery. Recognizing our inherent complexity helps us to maintain a sense of curiosity about our sometimes-conflicting internal voices. It is possible to "ask" different parts of the nervous system what they need, what they want to say or do. Are you excited to go out to a party (sympathetic activation), or do you need to rest (parasympathetic integration)? Do you want to buy tickets to travel abroad (cortical planning), or are you uneasy with flying (enteric)? Only when one part is constantly silenced do we find ourselves with serious nervous system compromise. Negotiations are one thing: "I'll stay up too late these three nights to finish my project, but on the weekend, I'll rest." Repression is another: never listening and silencing your body's needs.

Connections

Today you are invited to meet your nervous system as a friend and companion. Begin by noticing the current state of your body: do you feel relaxed, hyper-aroused, sleepy, introspective, or simply alert? Without judgment, shake out your spine and energize your central nervous system. At several times during your day, notice the interplay of outward excitement and interaction (through the sympathetic nervous system) and inner restfulness (through parasympathetic integration). Be sure to check in with your gut body as well as your brain. Sometimes they work together; sometimes one dominates the other. Inviting resonance through self-regulating a balanced nervous system is a foundational skill involving whole-body awareness.

TO DO

PRACTICING PRESENCE
3 minutes

Stress comes and goes during our days; rather than let stress drive our lives through nervous system hyper-reactivity, we can self-manage our responses, moving on and off center with resilience.

- Bring your chest forward, reflecting arousal in your body (sympathetic nervous system), like when you are late for an appointment or anticipating giving a public presentation.
- Soften your chest backward, reflecting a digestive and integrative state in the body (parasympathetic nervous system), like when you've just woken up or have eaten a big meal and are deeply relaxed.
- Now, bring the spine into balanced alignment. Notice how this feels in your body, being "toned" and available without anticipating or recovering—simply awake. This is an ideal state for cultivating presence in communication.

TO WRITE

DIFFERENTIATING THE NERVOUS SYSTEM 👁
10 minutes

Differentiating the parts of the nervous system helps us appreciate our complexity. This opens possibilities for resilience and responsiveness—and a sense of humor. With a writing pad in hand, begin where you are. Stay in one location so you can focus. Write for ten minutes, jotting down whatever outer and inner stimuli shift your attention. Include what part of the nervous system is most activated. For example:

- Your cell phone rings/dings—your midbrain is alerted.
- You hurry to answer before it stops ringing (sympathetic nervous system activation).
- You walk quickly across the room to find it (spinal cord reflexes).
- You feel irritated that it was just a telemarketer (limbic system).
- You settle into a chair and feel sleepy; your attention wanders and your stomach rumbles (parasympathetic nervous system).
- You slowly sense someone might be coming to the door of your room (enteric nervous system).
- A friend enters, and you check your watch to see what time it is (limbic brain and cortex).

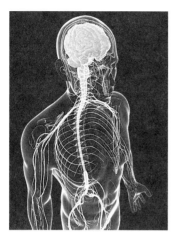

Brain and spinal cord
Graphic © Sciepro / Science Photo Library

TERMINOLOGY AND FUNCTIONS: SPINAL CORD AND BRAIN

Spinal cord: offers survival patterns and a two-way communication between brain and all other body parts.

Brainstem (midbrain, pons, medulla oblongata): a central pattern generator regulating breath, heart rate, and organ functioning as well as filtering sensory impulses.

Cerebellum: integrates and coordinates functions related to movement, balance, and posture; movement memory and other types of cognitive-emotional processes.[8]

Limbic system: hypothalamus, thalamus, and midbrain, plus a ring of additional structures: emotional-relational brain, important in memory.

- *Thalamus*: processes all incoming sensory information from the brain and other body parts; functions as a relay station from external inputs to higher-level cortical areas.
- *Hypothalamus*: governs the autonomic nervous system, which integrates emotional reactions with visceral reflexes; central pattern generator for the endocrine system.
- *Amygdalae*: Two almond-shaped clusters within the temporal lobes of the brain; regulate ability to feel and perceive emotions, including fear responses; notice emotionally relevant newness.
- *Hippocampus*: Principally involved in storing long-term memories; also allows spatial processing and navigation and keeping track of time.

Paired cerebral hemispheres (each hemisphere is divided into four lobes that work as an interconnected whole: frontal, parietal, temporal, occipital): govern decision making, planning, sensory and motor functions, language, and various aspects of memory; multiple associative functions (such as integrating sensory and motor experiences).

Corpus callosum: bridge for communication between cerebral hemispheres.

Dancing with the Endocrine System
Day 10

With Shruthi Mahalingaiah, reproductive endocrinologist

The endocrine system is a continuous communication loop. It involves our glands and their specific secretions that affect the whole body. There is no one place where it begins and no one place it ends. Mental and physical states may change almost instantaneously or over minutes, hours, weeks, or years through the regulated release of chemicals called hormones into the bloodstream. Paired with the nervous system through the hypothalamus in the brain, the neuroendocrine system oversees survival, well-being, and reproductive potentialities throughout our lives. Within energy-sustaining feedback-feedforward negotiations, the body's goal is maintaining homeostasis—balanced functioning in relation to outer context. While maintaining this "optimal state," endocrine glands give us the buoyant dynamics of a leap, the Eros of connection, and the deep rest of peaceful sleep.

A hormone is a chemical that is made by one cell to communicate with another cell or group of cells. Often referred to as "chemical messengers," hormones are found in all multicellular organisms, and their role is to provide an internal communication system between cells in close proximity or farther away. In humans, endocrine glands in the head, neck, torso, and pelvis communicate through the body's "internal ocean" as hormones are carried through the bloodstream. Communication occurs cell-to-cell and extends to the whole, just as you can talk to the person next to you or send messages to distant lands.

A healthy endocrine system functions largely without our conscious awareness. It comes to our attention when something goes wrong: no ovulation, sudden weight gain or hair loss, or excessive mood swings for example. These are signals to the conscious mind to attend! For preventive wellness and smooth functioning of all the body systems, it is crucial to address endocrine balance *before* problems occur. There are six areas that we can be aware of on a daily basis: *Water intake* (cells need water! One of the major requirements for good brain function is hydration). *Nutritional balance* (different tissues need specific nutrients along with a balanced diet, varying with body type). *Movement*, to keep the blood circulating and tissues toned (sitting for many hours at a time wreaks havoc with our bodies). *Breath*, to signal to the hypothalamus that all is well and to provide oxygen to all the cells. *Self-managing stress* (stress is not a thing, it is a perception; we can't touch it, but we can manage our mental attitudes, affecting health and longevity).

All the body systems are interconnected and multidimensional not only in the moment but across time. — Dr. Shruthi Mahalingaiah, in an interview

From Doing to Being
Whenever I give a speech or perform a dance, I know my endocrine system will be activated to enhance or deter. As any professional dancer, athlete, or innovator knows, you dedicate hours to rehearsal, training, and research in preparation, but the place of integrated flow and discovery happens beyond (and below) what your conscious mind can micromanage. Through the endocrine system, you are moving between the worlds of nervous system "intending" to endocrine "inhabiting"—from doing to being. When all is going well, you feel the flow.

David Thomson and Kathleen Fisher
in *Landing/Place* (2004)
Choreography and direction by Bebe Miller,
Bebe Miller Company
Photograph © 2004 Lois Greenfield

Conversations

"You close your eyes and put on an
imaginary pair of glasses," Dr. Shruthi
Mahalingaiah tells me on the phone.
"Your hands are along the side bars of the
glasses, near your ears. Right in the center,
between your hands, is the location of the
soft tissues of the head glands: the pituitary
and pineal glands and the hypothalamus.
We use hard skeletal markers like skull
bones as locators, but the glands themselves are luscious tissues." This is how she
directed a group of reproductive physiology students to experientially locate these
important structures.

The endocrine system is also affected by chemical messengers from
the environment. *Choosing products* to reduce exposure to endocrine
disruptors in our food, personal-care and home-cleaning products, and
maintaining our built environments (heating, ventilation, and air conditioning) to purify the air we breathe is important. We can recognize that endocrine disruptors (chemicals that can interfere with endocrine function)
are present in our contemporary environments. From the off-gassing of our
carpets and flooring, to chemicals in our lipsticks and face creams, as well
as sealants in plastic water bottles, we need to be attentive without amplifying stress. Noticing and listening to embodied, nonverbal responses to
environmental cues like skin reactions, breathing restrictions, or energy-
depleting allergies can alert us to the presence of harmful chemicals in
our midst.[1]

Endocrine balance can be supported by effective alignment, as the
bones help maintain dynamic space and blood flow between the paired
and single glands. Slouching can compress the heart and other tissues,
resulting in sluggish circulation and diminished interconnectivity of chemical messages. In dynamic alignment, when adrenaline-like hormones are
released in balance with other hormones, there can be a sense of vitality,
radiance, and expanded body beyond neuromuscular effort. For example,
during endurance hiking or running a marathon, an effective endocrine

bath is sometimes described as "flow state" or "being in the zone." You can see endocrine activity visually in athletes or dancers: the powerful fluid tennis serve of Serena Williams, the seemingly effortless gymnastics routine by Katelyn Ohashi, and the transcendent ballet leap by Mikhail Baryshnikov reflect pervasive endocrine support. Interruption of this unitive flow can result in over-efforting or scattered, unfocused energy. For fun, try jumping with your muscles; then, activate the endocrine system (see "to dos") and notice the difference in lightness and energetic output.

FACTS

Endocrine glands spaced throughout the core of the body communicate through the bloodstream; muscle tension and dehydration can interrupt the flow. Although glands are often studied starting from the top and moving downward, from an experiential perspective—because of their potency—it is important to ground and support the process of exploration by beginning with the lower centers and their bony skeletal landmarks. The word "lower" indicates that they are spatially below and evolutionarily earlier—not less important.

Starting with the pelvis and moving up toward the head, the following structures are currently recognized as belonging to the endocrine system because they release known hormones into the bloodstream: *reproductive glands* (two ovaries, testes); *adrenal glands* (two glands, located atop the paired kidneys); *pancreas* (below the breastbone, about six inches in length, extending from the front to the back of the body); *heart* (behind the sternum, between the lungs, reclassified in 1983 as an endocrine gland because it secretes a variety of hormones);[2] *thymus gland* (above the heart behind the upper sternum/manubrium); *thyroid gland* (two lobes shaped like a butterfly, located at the front of the neck); *parathyroid gland* (two glands, located on the posterior surface of the thyroid); *pituitary gland* (two lobes, suspended by a stalk from the hypothalamus in the brain); and *pineal gland* (single structure, tucked in a groove between the two halves of the thalamus at the midline of the brain). The kidneys, digestive tract, and placenta have known endocrine functions as well. Every cell in the body has the capacity to respond to endocrine signaling.

Along with the endocrine glands described above, Bonnie Bainbridge Cohen, in her experiential work through the School for Body-Mind Centering, includes groupings of cells called *bodies* in her overview of the endocrine system. The *perineal body* (in the center of the pelvic floor), the *coccygeal body* (a small cluster of cells at the interior tip of the coccyx), the *carotid bodies* (on each side of the neck along the carotid artery), and the *mammillary bodies* (paired structures—named for their similarity to small breasts—on either side of the midline of the brain, on a diagonal axis between the pituitary and pineal glands) are explored and described. These little tissue clusters are classified as bodies because isolated measurable secretions have not yet been discovered; they are, however, highly vascularized, which suggests that they are likely to secrete hormones into the bloodstream.[3] At an experiential level these tissues have characteristics of

Metabolic Medley

When Shruthi was pregnant with twins, she experienced gestational diabetes. As a doctor, she knew what to do: an early morning, pre- and post-meal finger-prick test to determine blood sugar levels and calculate how to balance them by selecting the appropriate foods. I watched her eat greens with her scrambled eggs and toast in the morning and take a walk after a carbohydrate-laden meal to keep the numbers in range: this made all the difference in neutralizing the high blood sugar levels that can result after eating carbohydrates. She shared these practices with my husband and me, and it made all the difference—we shifted our sugary desserts to after lunch and then took a walk after dinner instead of having to digest an enormous meal. Sharing that information with family and students helped others stabilize energy levels and curb weight gain. Simple information can have big effects!

Drawing by Avani, age 9

Endocrine glands and related organs
(Note: In a three-dimensional living body,
the stomach and pancreas are above and in
front of the kidneys; see Day 25.)
Anatomical illustration by Nancy Haver (2021)

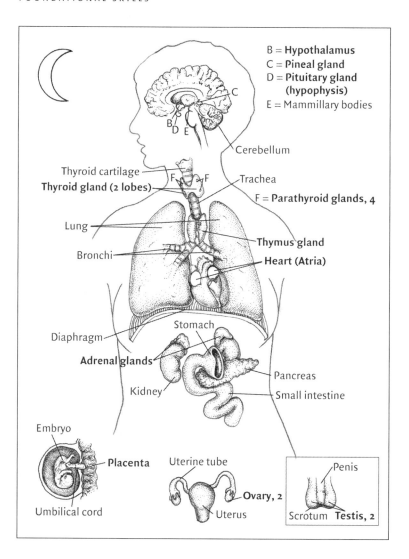

B = **Hypothalamus**
C = **Pineal gland**
D = **Pituitary gland
(hypophysis)**
E = Mammillary bodies

Cerebellum

Thyroid cartilage

Thyroid gland (2 lobes)

F = Parathyroid glands, 4

Trachea

Lung

Bronchi

Thymus gland

Heart (Atria)

Diaphragm

Stomach

Adrenal glands

Kidney

Pancreas

Small intestine

Embryo

Placenta

Umbilical cord

Uterine tube

Ovary, 2

Uterus

Penis

Scrotum   **Testis, 2**

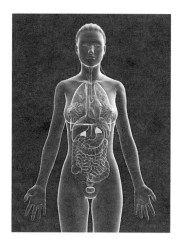

Healthy adrenal glands
Graphic © Sciepro / Science Photo Library

the endocrine system. By exploring all the glands and bodies for yourself through sound and movement, you can notice what you find. The totality of glands and bodies corresponds roughly to the seven major chakras in yoga.[4]

APPRECIATING THE HYPOTHALAMUS AND PITUITARY GLAND

The *hypothalamus* is considered the keystone of the endocrine system, with regulatory responsibilities including energy balance, water balance, growth and development, and reproduction. Located in the center of the brain, the hypothalamus is a place of communication between one system and another—neural with endocrine. It links the autonomic nervous system to the endocrine system, modulating between fight/flight/freeze sympathetic activation and integrate/digest/rest parasympathetic relaxation in response to perceived stimuli. Signaling can come from cell to brain and brain to cell. For example, when sensory signals in the nervous system detect danger, the endocrine system activates the necessary chemical ingredients for response: adrenaline might be released from the adrenal glands. The nervous system perceives and records, and the endocrine creates whole-body tonal responsiveness. Pressing your tongue to the roof of your mouth and drawing a line up and back will help get a sense of the location of this important region in the center of the brain that is also home to the pituitary, mammillary, and pineal glands.

The hypothalamus, as the keystone or intermediary, partners with the attached *pituitary gland* to link to the endocrine system. Communication from the hypothalamus to the pituitary occurs through synaptic pulses, modulating amplitude, frequency, and duration to specify response to the body's needs. The pituitary is small but mighty. Slightly below and connected to the hypothalamus, this pea-sized gland with paired lobes secretes all of the *stimulating* (*releasing*) *hormones* that act at endocrine gland sites in various parts of the body. Target glands, such as the thyroid gland, adrenal glands, and gonads respond by secreting *final hormones* (such as thyroxin, cortisol, estrogen, and testosterone) that both activate and feedforward to target tissues, and feedbackward into the loop. If you are under perceived stress, for example, reproduction may be put on hold as the body prepares for defense or conserves energy for survival. Because this unique area of the brain is considered a "pulse generator" and "integrator" throughout our lives, aging and sexual responsiveness are implicated.

Most hormones are modulated by a sensitive feedback loop. When the *final hormone* is synthesized and in sufficient supply, the concentration of the *stimulating hormone* is often reduced. When there are low circulating levels of a *final hormone*, there is a resultant increase in the *stimulating hormone* to achieve homeostasis in the system, so the organs are functioning optimally. Hormones have different lifespans in the body: For example, the adrenal response is meant to be fast, giving a quick-burst or surge of energy; yet its impact lasts up to an hour (and you can feel burned out if the adrenals are stimulated too often or for too long). If the fight/flight/freeze response is activated, the body needs about twenty minutes to physically recover. A sugar high that registers in your adrenal and thyroid glands lasts about forty minutes before the sugar crash takes over. It takes about five days of adding physical activity like jogging or vigorous dancing into your lifestyle for the bones to respond by laying down more calcium. The rhythmic pulsatility of the hormone-releasing neurons in the hypothalamus impacts sexuality and varies between individuals. Generally, it increases at night, fluctuates across a lifespan, and most notably, is "on" during puberty and the reproductive years. When we add medications and recreational drugs into this mix, we impact the hormone balance. There is always more to learn about the mysterious diversity of hormones and endocrine tissues and their impacts on the human body.

We can understand hormones through associations with past endocrine responses. *Adrenaline*, for example, is released from the adrenal glands atop the kidneys when you are startled by an odd sound (Oh!) or watching an exciting sports play (Yay!). *Oxytocin* from the posterior pituitary gland is released in heartwarming moments of bonding or cuddling, like holding a newborn baby or nestling into a place in nature that you love.

Wake-Up Call

My sister Karen taught me about the "4 a.m. worries." She described waking up early in the mornings with all the concerns and unprocessed thoughts from the day before. Things you didn't even know were upsetting or unfinished are there waiting for you: the missed item on the shopping list, the phone call, what someone said. Reading that there is a cortisol increase or "bump" between 3 and 5 a.m. clarifies: you get an energized dose of activation at that time. I wonder what the evolutionary advantage might be for early morning (sleep-disturbing) wakefulness? Now I say to myself, it's okay, greet the rush of urgency with some calm attention, and return to resting. And if that doesn't work, get up and write down your thoughts, then enjoy the sunrise.

Baby's Breath

Once during travel, I spent the night with a friend and met her newborn. As a new mother she was so exhausted in the early morning hours after nursing that she put baby Caitlin on my chest so she could sleep. What a shower of sensations! Intimacy, so sweet, feeling that young being breathing, heartbeat to heartbeat, all twelve pounds supported and growing. A surge of connection (oxytocin in action) flowed through my body, bonding. Years later we met in Paris, our hearts greeting each other in their own knowing ways.

Endocrinized States

Some events stay with us forever. Living in Paris in my twenties, I saw the amazing Russian dancer Rudolph Nureyev perform at the Paris Opera. His leaps were so spectacular that I looked heavenward and there were cherubs painted on the vaulted ceiling—ecstasy! Merce Cunningham could be instantaneously in the air when dancing with no trace of preparation or effort. A few students have that buoyancy as well—they are most at home air-bound. Knowing the effort it takes me to create a bounding leap, I am fascinated. Studying the endocrine system with Bonnie Bainbridge Cohen, we activated glandular support—particularly the coccygeal body and gonads. There I was mid-air, seemingly effortless. Every system in the body has its impact on movement; this one adds a spectacular edge.

When I was asked to officiate at an outdoor wedding, I felt honored and concerned—what does that role involve? During the ceremony, I stood on a rock to induct the couple through that ritualized ceremony. As I invoked the vows, I felt my body shaking. I wasn't nervous, what was going on? The energy charge was adrenalized, but not from fear. There was no option to move like when performing, riding excitation into expression. Just the voice. It felt like a volcano erupting; more energy than I could channel—exciting (although a bit worrisome to watchers).

Now I recognize those moments when someone is in an "endocrinized" state. You may as well not try to talk them out of it or explain what is happening. It takes time for hormones to flow to their target sites and also to leave the system. We can learn to enjoy or tolerate the sensations in ourselves and others—and decide how, when, and where to say "yes" or "no" to that condition.

T-lymphocytes are created within the thymus gland to fight infection when you are feeling sick or perceive threat. *Thyroid hormones* (T_3, T_4) can result in hypothyroidism (too little): you feel cold and tired, lose hair, have dry skin; or hyperthyroidism (too much): you may feel hot, have tremors, lose weight. *Melatonin* is secreted by the pineal gland in the brain in response to darkness, making you sleepy. *Testosterone* and *estrogen* from male and female gonads are vitally linked to every tissue in our bodies, affecting mental cognition, tissue elasticity, muscle growth, and bone density, as well as the health of the eggs and sperm. These hormones, among many others, work together as well as independently to maintain inner equilibrium. There is more to this complex system, but personal experience can provide an embodied baseline for exploration.

When things are out of balance, our attention is drawn—sometimes dramatically—to this important system. Focusing on what creates optimal health and wellness can be a priority even though it is harder to measure positive, life-affirming responses than negative impacts. New brain-imaging techniques are helping us "map the mind," showing the effects of positive thoughts on endocrine activity as well as the importance of movement and mindfulness techniques to create balanced conditions.

JOINING THE DANCE

Practicing attentiveness in our daily lives—playfully "dancing" with the endocrine system—requires tuning in. Rather than dominance over our body, we find optimal flow. We know from social dancing with a partner that there is give and take, a constant negotiating of timing, rhythm, force. If one partner dominates, there is conflict and awkwardness, but when you are responsive and flowing together, there is an ever-shifting dynamic balance. Then you can attend to dimensions beyond the basic steps, including expressive qualities and communicating with others, enjoying the radiance that comes from pleasure and moving. Maximizing our human potential through the endocrine system, like partner dancing, requires skillful attention.

Research shows the benefits of movement, music, and meditation practices on hormonal balance. The soft, luscious tissues of the endocrine glands are particularly responsive to resonant sound vibrations that increase circulation to the region. You can touch specific areas of the architecture of the body, like the breastbone, front of the neck, or skull to identify the location of endocrine glands while bringing vibratory sound to these areas (*ah, oo, ee*) at different pitches—not to overwhelm but to find ease. Bony skeletal landmarks help focus attention and stay grounded as you explore.

Communication feedback loops are the hallmark of the endocrine system. Responding to sensory signals from the outside world, as well as from the internal environment of the body, the endocrine system supports all critical functions for survival and for enjoyment. While maintaining a balanced state, the glands bring excitement and connectivity into our lives. They underlie the capacity for heightened states of consciousness and intuition as well as dynamic physical responses. This powerful and complex system, in partnership with the nervous system, is our natural inheritance. Noticing and "dancing with" this multi-dimensionality, rather than resisting or trying to dominate, enhances our lives individually as well as with families, friends, nature, and our global community.

Connections

The invitation today is to notice and value any times when you experience feeling relaxed and internally balanced, practicing inhabiting the "body of your life" more spaciously. If you are used to living in an adrenalized state or with stress, depression, or anxiety, it does not always feel good to listen inwardly. It will be helpful to invite a playful and self-respectful attitude into your day to boost your mood and encourage more inner radiance and flow. Revitalizing endocrine activity in the places that may have closed down or been "propped up" physically or emotionally in the past is an embodiment skill for living and communicating with more internal balance and ease.

TO DO

ELEVATED PELVIS (MODIFIED BRIDGE POSE IN YOGA)
10 minutes

When locating the glands, it can be helpful to feel the support of the ground; relax and turn the senses inward.

Lying in constructive rest, eyes closed:

- Lift your pelvis off the floor so that it creates a long diagonal from your knees to your shoulders. Keep the neck free, chin slightly toward the chest as though a river is running down your core and through your head—no restrictions.
- Breathe deeply, allowing your organs to relax toward your chest.
- Lower your spine slowly, part by part (vertebra by vertebra), elongating your pelvis toward your heels. Be sure the knees remain parallel to each other.
- Repeat, lifting your hips slightly higher and allowing weight to rest on top of your shoulder girdle without compressing the neck.
- Visualize the endocrine glands in your head, supported and resting.
- The thyroid and parathyroid glands along the front of your neck are particularly stimulated by this inverted pose.
- Now, as you lower the spine, release the glands along the front of the spine toward gravity: the thymus (upper chest), heart (center of the ribs), pancreas (below ribs), adrenals (back of floating ribs), gonads (ovaries in women, testes in men).
- Release your tailbone (coccygeal body) and pelvic floor (perineal body).

Obsessions

Bird calls are everywhere this June in Maine. I am normally overseas in this season, so I'm captivated by this seasonal behavior: arrivals, territorial battles, nest building, nestlings, and feedings—all hormonally orchestrated for successful reproduction and survival. Author Julie Zickefoose is a true obsessive: a bird rehabilitator and visual artist who has gone to extremes to know and help birds since childhood. Reading her stories, I now watch birds with a more nuanced eye, recognizing patterned feeding behaviors, the calls of young that are demanding to be fed every twenty minutes for two weeks! On a nature walk identifying migrating warblers, our experienced leader tells us that on July 6 in this region of Maine, the birds stop calling. The power that seasonal change has on their small bodies is reflected in our own.

Surrendering to Flow

The stream in front of our home has a strong current. On a hot summer day, I submerge and yield to floating. This requires releasing "control" and allowing myself to be moved. There's a death-ness to this moment: yielding will to surrender. How long can I tolerate suspension, and at which moment will I put my feet down on the gravelly bottom and take charge? This embodied experience of outer flow informs inner flow. Fluids need to keep moving rather than be held—controlled, they become stagnant.

Wildness

Bonnie Bainbridge Cohen helps illuminate the systems: "There is a wildness to the glands and a sense of control to the nervous system."[5]

Miguel Castillo, Constructive Rest
Middlebury, Vermont (2015)
Body and Earth: Seven Web-Based Somatic Excursions
Still image from videography
by Scotty Hardwig

- Lying in constructive rest, allow your back body to support the organs and glands.
- To end, stretch in any way that feels good, extending outward through the periphery. Engage all your muscles and fascia, linking your center to fingertips and toes and circulating flow. Yawn and stretch the face, stick out your tongue, invite wholeness.

NOTE: In yoga, this inversion is called *setu bandhasana*, bridge pose. If you have difficulty articulating the parts of the spine, touch the front surface of your body as you move part by part. For description of constructive rest, see page 171.

TO WRITE

AN ENDOCRINIZED EXPERIENCE
10 minutes

Sometimes we are taken over by our own energies, and we simply feel "in the flow"—for better or worse. Write about a time when you felt taken over by an energetic, endocrinized state in your body. It might be a time in sports when everything felt synchronized within your team or an encounter that left you in an enraged whirl or cloud. How long did it last? What did you learn? Has it ever happened again? Can you feel it now?

III
MODES OF
COMMUNICATION

Nancy Stark Smith and dancers
in the *Underscore* (2007)
Helsinki, Finland
Photo © Raisa Kyllikki Ranta

Movement and Touch

Movement is our first language. It is how we come to know ourselves and the world. As babies, we move to feel ourselves moving, learning the boundaries of self and mapping the territory of our bodies in relation to people and place. The tongue, mouth, and eyes join the fingers and toes, exploring and identifying what we find. Essential muscles and organs develop, supporting our big heads and eventually propelling us through space. Exploratory gestures and vocalizations become words, marks, and meaning-making tools shaped by family, culture, and environment. This early history of our movement life resides in our bodies, informs our days, and underlies our communicative range.

Movement is multidimensional. It activates every breath, step, and gesture, including the inner flow of fluids throughout our tissues and outer expressive communication with our fellow humans. A whole range of possibilities accompanies us throughout our days, occurring at both conscious and unconscious levels. Every change of attention is part of our movement life as we redirect focus and orientation. Expressivity through the muscles in our bodies is part of human connectivity and refinement, affecting how we literally move through the world, hold ourselves, and exhibit postures, habits, and learned patterns of movement. Qualitative muscular range communicates our intent: rocking a baby quietly by our heart or curling our fists in frustration, we recognize each other through our movement and through the ways we utilize space, energy, and touch. Feeling the primacy of movement and its vital contribution to self-knowledge and culture, we can bring attention and focus to understanding this multilayered aspect of our being. It is the basis of all other modes of communication.

Valuing parts of ourselves that might not get acknowledged in restrictive social contexts can be refreshing. As we learn to listen to the language of the body, we can choose when and how to engage more consciously with movement qualities and impulses. Dancing and engaging in other forms of free movement as well as learning ancient and contemporary codified movement practices lets our bodies move us, allowing us to self-manage stress and increase our empathy for others. Part of injury and illness is the conflict between what we tell our body to do and what it needs to do for safety, healing, and easeful interactions. Clearing stuck movement patterns often requires allowing the unexpressed to be expressed in a supported context, broadening possibilities. Within our movement life and our experience of living and communicating, we need to be comfortable in our embodied selves. Stagnant energy drains our capacity for engagement, locks in negative moods, and limits our human potential for connection. As we open our movement range, we enlarge possibilities for both receptivity and expressivity through our moving body.

The language of movement is our most fluent tongue and a language all human beings hold in common. — Anna Halprin, from Making Dances That Matter: Resources for Community Creativity

Fishing for Life

We are fly-fishing in Grand Lake Stream, Maine, across from our house. It is early morning, before black flies arrive full force. This is a family tradition: my husband and stepsons are excellent fly fishers; for me it is mostly an excuse to spend long periods standing in fast-moving streams. The power of your cast comes from your feet—and they are down there underwater. While your eyes see one thing, your feet know another, seeking stability on often-slippery rocks. After casting and recasting, awaiting the tiniest tug on your line, there's the moment of connection—fish to you. Holding those streamlined, landlocked salmon in our palms underwater, feeling their muscular heft and aliveness, watching their wildness swim free when released is somehow exhilarating—at least for the human. Both sons have made a life with fish: one is a biologist doing salmon habitat restoration and the other sends regular missives from backcountry fishing. This lifeline between us humans and fish, father and sons, is slippery but constant. Those underwater fish are—as we know—part of our lineage, our spinal ancestry, our hope.

EMBODIED COMMUNICATION

Exploring movement as a component of embodiment, we may face basic fears and internalized judgments around our body: fear of stillness, fear of being alone, fear of being empty inside, fear of not being loved. "I'm too fat. I'm too thin. If I'm not moving, I don't exist. If I'm not seen, I'm nobody. If I don't do something good or meaningful, nobody will love me. If I'm embodied, I'm not spiritual." Although these statements may seem harsh, they are commonly expressed in movement classes and experiential sessions. These unconscious blockages affect what we believe to be possible and what we believe ourselves to be worthy of in our lives and interactions. As we close our eyes and listen to our bodies there is also the potential of accepting ourselves just as we are.

Replacing inhibition with inquiry, we can learn how rich our movement life really is. Generally, we push into the body's unconscious what we consider to be negative—our sadness, our anger, our fear. But below that layer of unexpressed movement is a wealth of human experience. That is the resource we want to bring with us as embodied communicators. In this process, the body is our guide. Our entire life history as well as foundational movement patterns are stored in our bodies—memories reside in our tissues, not just our brain. We can learn to listen to the stories and intelligence within. As we recognize movement as a resource for interconnectivity within our skin, we develop a dialogue with our multilayered self.

Beyond functional gestures and actions, moving connects us to three layers of *personal*, *collective*, and *transpersonal* dimensions that arise from the body. When we explore movement within the experiential practices in this chapter, or an eyes-closed form like Authentic Movement (see Day 28), each layer is amplified in specificity. The *personal movement* layer may include one's present-moment need to refresh, heal, or express. This could be the impulse to stretch and circulate energy to release tension, to lie down and rest, or to simply feel and enjoy, bringing more awareness to what we are doing. Healing injuries may be involved; sometimes our bodies know how to unwind compressed nerve tissue even more efficiently than a specialist because the knowing comes from inside. We can utilize the body's wisdom for self-healing and integration.[1]

Collective movement includes what we are picking up from others. We are influenced by those around us whether we intend to be or not. This can involve physical cues from other bodies or the tone of a gathering or room; something broader like collective holidays, cultural disruptions, or celebrations; or events occurring in the world. We are impacted by an intuitive awareness of what is happening in a space or place. If someone is particularly noisy or loud in a group, for example, you may become silent to balance the energy field. Conversely, you may join the expressivity, or you may directly resist it. Often in an eyes-closed group of movers within Authentic Movement, a gesture is passed around. One mover extends an arm toward the sky and others in the group do the same movement without any visual cue. In more daily/public group settings, if one person is quite tense with a rigid spine, others tighten. If someone pauses to stretch, others feel free to move. If there is a general atmosphere of relaxation, it impacts expressivity. Awareness of the collective layer of movement lets us discern what is ours, what belongs to someone else or the group, and how to embody and clarify our intention so we can interact most effectively and authentically.

Transpersonal movement is another, more mysterious but ever-present layer. This involves archetypal gestures or postures beyond personal or collective intentions and influences that arise from the body. Gestures of offering, prostrations, or trance-inducing patterns like rocking or spinning can emerge from intuitive realms beyond our specific heritage. Hand mudras that we have never been taught may surface distinctly at this level. When moving eyes-closed or in other receptive states, we may open to transpersonal energies beyond our awareness and rational understanding, transcending cultural boundaries and historic timelines. Recognizing these natural and insightful connections helps us feel part of a larger interconnected world, expanding limits of time and space through awareness of the multiple dimensions of our movement life.

REFINEMENTS

Inner experience and outer form may need some tweaking to be congruent. A teacher, mirror, or video camera can reflect back what is being seen while you are moving and feeling, connecting inner experience with outer feedback. Allowing oneself to see/feel an experience helps one develop a more conscious and discerning inner guide to bring visible movement expression in line with intentions. When the body and psyche are in sync, there is often a kind of radiance that is apparent. In this process, the body and the unconscious self, or psyche, develop a dialogue with the conscious self—a pact of trust that is constantly being negotiated. For example, in an injury of the spine, the muscles spasm to protect us from movement. (If we didn't hurt, we would move!) As the body trusts that we will listen to its needs and rest, the spasm can release. That is part of the negotiation. Can we trust ourselves?

LIMITATIONS

Inhibition is part of efficiency. It allows us to sit in a room with others and listen, take turns, and respond effectively. Inhibitory pathways in the brainstem smooth out movement impulses and sequences, trimming away extraneous bits. Too little inhibition can result in trembling, fidgeting, or spasticity. But too much inhibition limits potentials and increases stress. In some situations, we may want to speak up, run away, or throw a fit in response to what is happening, but instead we stay within appropriate boundaries, whether accurately perceived or imagined. Yet this inhibition gets stored in our nervous system, muscles, and fascia as uncompleted impulses and can remain as stress, tension, and restrictions that deplete our energy. It takes a lot of work to hold everything inside. Or we may create a situation where that expression is completed elsewhere out of context where it is potentially destructive, like yelling at children or punishing a

Endurance
Teaching with Wangari Maathai at the Nobel Peace symposium in Iowa, I found her presence formidable. Over lunch, we talk about endurance and the demands of her schedule. She faces challenges in her environmental activism, including opposition and death threats for planting trees with women's groups to reforest the land. When I ask how she sustains her schedule, she tells me she works out every morning: "Wherever I am, even if it's 4 a.m., I exercise. That sustains me." Strength isn't just in words, it's in your whole being. After her keynote speech and a panel, she attends my embodiment workshop. She enters, lies down right at my feet in the crowded classroom, closes her eyes, and yields her weight to the earth. I look at her dry and cracked soles, which have seen many miles. She teaches me humility, as well as powerful presence. They go hand in hand.

Rehabilitating Touch

Dancer Felice Wolfzahn and I are traveling by train through Germany on our way to teach a workshop in Italy. She tells me of her grandparents' escape to Brazil, then to Canada, from Nazi Germany. Part of her interest in offering contact improvisation (CI) internationally is rehabilitating intercultural trust through touch. Contact is an improvised dance form where you share a point of skin contact and follow that place of connection. It "is based on the communication between two moving bodies that are in physical contact and their combined relationship to the physical laws that govern their motion—gravity, momentum, inertia."[3] These unconscious reflexes are more universal than personality or culture: it's a dance with physics. CI is practiced around the world, often with groups who would be considered "in conflict"—Israelis and Palestinians, Turks and Greeks, etc. We look for what we have in common and move from that place.

dog. Finding appropriate, useful, consistent practices to clear blocked energy in the body helps flush toxic buildup and release fixed postural habits, allowing us to rechannel and refuel energy in easeful and appropriate ways.

Movement limitations can be lodged in any system in the body. If we hold the idea that the body is a machine, the limitation can be in our thoughts. If our muscles aren't toned and responsive, the fascial system that covers and wraps all the tissues can be compromised and stiff, restricting nuanced responses. If the bones are not supporting our weight, the organs can be compressed. To refresh movement, we can work through any system to impact the interconnected whole. Kinesthetic literacy lets us notice how we move in the moment, offering insight into the conditions and context for our movement choices.

Every person has a richly detailed, multilayered movement life. Yet these dimensions may not be visible or acknowledged unless specifically invited and valued in a safe-enough context. Bringing consciousness to these experiences, rather than dismissing them or overly exaggerating their uniqueness, lets the extraordinary be ordinary. Movement dimensionality is part of our human nature and our responsibility—it is our capacity to respond to the moment at hand. As we breathe, converse, or hug in our daily lives, we can inhabit ourselves with more ease and authenticity, regardless of circumstance.

TOUCHING AND BEING TOUCHED

Touch underlies all other senses, providing accurate, continuous information about self and the environment for safety and survival. The fact that touch is essential to physical, mental, and emotional health is repeatedly documented in studies with both animals and humans. Statistics in orphanages show that touch deprivation affects development and can result in retardation or death. Premature babies in hospitals develop 50 percent faster if given loving tactile stimulation. Research conclusively links violent crime with deprivation of physical pleasure and harsh childhoods, including trauma at birth.[2] With all these statistics showing that touch is literally food for the body, why are many people starved for healthy touch? Instead of appropriate models of touch, we are continuously fed images linking touch to sexuality and violence in the news, entertainment media, and advertising, which often arouse a confused mix of fear, desire, shame, or insecurity with the goal of capturing our attention and selling us things! But our bodies recognize and thrive on real touch, real intimacy. We can learn to value what is inherent in our intelligent moving bodies.

As we touch, we are also being touched. In this two-way conversation, millions of sensory and motor nerve endings are involved, registering and responding to input. Our feet touch the ground, and we receive information about location; our skin touches the air, clothing, and other body parts, which in turn inform our sense of body position in space. When you touch someone, you are not just doing something to someone else, you are being touched and affected in the process. Conscious touch with another person is *consensual touch*: there is an unspoken or spoken agreement toward exchange that requires discernment.

Practicing intentional touch involves awareness of layers: If you place your hand on your own shoulder, you can feel the cloth. What did you wear today? Go a little deeper, and you can notice the skin. You are not touching it directly, but you can tell when you are touching that sensitive, resilient layer of the body. Then you can go a little deeper, into the meat of the arm— your muscles; they are generally quite ready to be rubbed. Attending to the muscle layer, you can slide your hand all the way down to your elbow, where you might notice the tough fibrous tendons, attaching muscles to bone. In that area near the joint, it is easy to feel the bone.

Attending to touch requires focus. As you reverse your sequence, return to muscles and notice their textures, back to slippery fascia surrounding muscles and linking to skin as outer boundary, then to cloth. When you remove your hand, you can feel heat in the space between hand and body: the electromagnetic field—measurable and real. In this process you are becoming aware of touch specificity by feeling the different layers and depths. Each level of touch is useful for treating different conditions or addressing particular communication scenarios. Just standing next to someone affects what happens; cloth level can be quite appropriate when working with someone with extreme touch sensitivity, or for whom skin is vulnerable—like touching an elderly relative. Bones are the architecture of the body and can benefit from effective, knowledgeable touch to redirect and realign. The point of differentiating layers is to be touch specific; clear in your intention and communication before, during, and after touching and being touched. Throughout, we remember that touch is reciprocal exchange.

Like any good communication process, listening is involved as well as timing, absorption, and reflection. When facilitating embodied, experiential explorations with yourself and others, you have to slow down to allow time for movement and touch to register throughout the body. The inquiry is how to go beyond the idea or image of an experience, resulting in just talking or thinking about it. Instead, you clarify the idea and image of what you are going to do (visualization); you try it (somatization); and repeat the practice over time for absorption and integration (embodiment).[4] Once the pathways are known, the body can respond by moving you—the nerve networks, kinesthetic understanding, and cellular memories are established. In the "to dos" in this book, for example, once you have engaged a sequence in depth, the process moves more quickly. You can orient yourself to ground, space, and breath at various moments in your day because the pathways are already established.

Connections

For these next hours, encourage more movement at various intervals in your day. For a break from focused intensity, rather than eating your favorite treat or turning to an automated device for distraction, nourish yourself

Ceramic Box 0706 (2006)
by Harriet Brickman
Raku-fired clay
Photograph by Anya Brickman Raredon

Dancing

At a memorial dance performance in honor of my mother held at Millikin University, where both of my parents attended and taught, there was a glass display case in the foyer that we filled with old photos of their college days. As I stood gazing, a woman came up to me and said: "I am still in love with your father. He was the best dancer—all the girls wanted to dance with him. Is that why you are a dancer?" Hmmmm . . .

with five minutes of moving—a "movement snack." Get up out of your chair and shake out your legs; find a quiet corner and stretch to the sky; touch the ground with your hands, release your head and neck, and notice the impact on your productivity and energy levels. Movement activates blood flow through tight muscles, restores and refreshes expressive range, and supports resilience in communicative possibilities. Once you overcome the inertia of beginning, you know you'll feel better.

TO DO

KNOCKING ON THE DOOR OF LIFE (QIGONG) 👁

5 minutes

Activating sensations in the body through movement and touch helps refresh vitality and remove tension. This spinal enlivening exercise pumps fluids in and around your spinal column.

Begin in basic standing posture:

- Begin rotating your spine, turning from side to side from your hips and waist.
- Let your arms flow freely around your body with this spiraling movement (like a flag wrapping around a flagpole).
- Allow your hands to knock across your abdomen and lower back, one in front, one in back.
- Repeat several times until this knocking movement is easeful and natural. This stimulates key pressure points for energy and vitality and loosens the lower back.
- Breathe deeply and find just the right amount of pressure for your "knocking" to wake up energy in the body.
- Now, make light fists with your hands and knock on the chest. One hand knocks just below the collarbone on your top ribs and the other on your lower back.
- Repeat several times. Keep your shoulders relaxed, unwinding from your center and changing which hand is in front with each twist of the spine. This activates energy in your lungs and stimulates the immune system.
- Let the whole spine be involved as you turn right, then left.
- Now, cup your hands and come up over your shoulder and neck to knock. One hand clears tightness out of the upper back and the neck, and the other continues to knock across your lower back with each turn.
- Repeat several times. This encourages blood flow up to your brain and helps release shoulder tension.
- Now, come back down to your chest; continue swinging the arms with ease. Repeat several times.
- Come back down to your low back and belly. Repeat several times.

- Slow your movement down, looking over each shoulder as you turn, getting a stretch through your neck and eyes.
- Relax and let your spine unwind into stillness.
- Soften your gaze or close your eyes, and notice any place where you can feel energy flowing in your body. This sensation may feel like heat, tingling, or pulsation in the tissues. Just notice as you become familiar with sensing energy moving through your body.
- Now, be brave and explore! Continue with eyes closed or with a soft focus, and follow your body's desire for movement. The set practice is an entranceway; the resulting exploration is your own movement excursion.

NOTE: These qigong practices are drawn from and approved by Lee Holden. For more information and video resources on Holden Qigong, see www.holdenqigong.com/.

TO WRITE
LAYERS OF TOUCH
10 minutes

Touch can be a challenging topic in many social situations. Let's take time to notice intentionality while touching in nature. Journey outside and locate a tree for a touch conversation. Rub your hands together to stimulate heat, then place one hand on the bark of the tree: press through the skin layer of your hand, the muscle layer, and connect the bones of your body to the stability of the tree. Lean forward and feel the support all the way down to the roots and earth. Shift your weight side to side and arc your awareness down through your own structure to your feet and the ground, creating a touch conversation. Now, return your weight back to self-supported standing. Notice the bones, muscles, skin, and heat of your palms as you release connection. Let this investigation invite writing about any experience of touching and being touched.

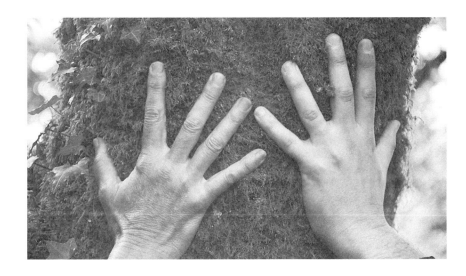

Hands and Tree
Pen Pynfarch, Wales (2015)
Body and Earth: Seven Web-Based Somatic Excursions
Still image from videography
by Scotty Hardwig

Day 12

Speaking and Listening

The quality of the expression always depends on the quality of the listening and reception.— Hubert Godard, from Reading the Body in Dance: A Model

Noises

My younger stepson was mostly silent; his brother did all the talking. So he created "animal noises," irritating little squeaks and beeps that would drive everyone crazy, his own language consistently inserted into conversations. This all exploded in high school when his brother went away to college. Breaking through, breaking out, breaking things, he found his voice. He told me I was a bad listener and that I exaggerated. He was right on both accounts—I get carried away with a good story. Looking back on that time, it was a transition. So much was kept inside, seeking a way out. On the phone now I leave long pauses, waiting for the next theme to emerge. Communicating is making space for feelings and thoughts to find form.

The dialogue between speaking and listening is a somatic exchange. We feel what we are saying and hearing, as we are literally touched by sound. Air is real. Sound waves travel and vibrate tissues. In effective communication, we respond to the immediacy of speaking and listening rather than preplanning what we are going to say or interrupting others with our ideas. In this process, there is a difference between "performance" and "preformance." We remain responsive to the shared moment at hand through effective listening skills rather than anticipating outcomes.

Hearing and listening are distinct processes. Hearing is a sensory activity—the ears house sensors, receiving sounds from the external environment and from the body. But listening is not only a sensory experience. There is a motor aspect involved—you have to focus your attention on specific aspects of what you are hearing. *This* sound, not *that* sound, gets prioritized. Intention and values play a role, involving emotional layers of interpretation and filtering. With individualized "selective hearing," your friend's voice gets through in a crowd and other voices do not.

If you are listening well, the pause or the space between focused listening and clear speaking allows absorption. During the "taking in" process, incoming data registers and is integrated through the brain, affecting breath and heart rate. The whole body is engaged, informing effective speaking. The quality of listening enhances possibilities for effective, multilayered communication that is connected to memory, emotion, and imagination.

FACTS

The ears are always open. The whole body listens. You are affected by sound at every moment, even when sleeping. Sound is vibration; sound waves register in sensory organs in your ears and are interpreted in the brain. Sound also resonates in all the body tissues. Bones, in particular, vibrate; you are touched by sound, regardless of whether you are consciously listening. Sound tells you where you are and also takes you far away through your imagination. The temporal lobes of the brain (around your ears) create detailed three-dimensional maps of your surroundings through sound to orient you to context moment by moment. They also link awareness with memory of prior places and sounds for interpretation and response. Locating sounds through audio maps, you understand the proximity of a fan, car, or person without having to see them.

Sound has emotional relevance. Although words must register in the cortex of the brain for meaning-making and interpretation, nonverbal sounds link directly to (and stimulate) the limbic system—the emotional, relational brain (see Day 9). The amygdala—the emotional sentinel of the limbic brain—registers anything new or potentially threatening. A crashing

Tea Couch (2006) by Kristina Madsen
Bubinga, dyed pearwood veneer,
ebony, linen
Photograph by David Stansbury

sound while you're sitting in a meeting generally activates your amygdala. Scanning sounds for survival implications affects body-level response and cues communication. Some cultures, families, and individuals are highly attuned to sound, especially in contexts where visual data is too complex and doesn't offer sufficient cues, like in dense jungles or cities.

If sounds are familiar, you can continue about your business. If the sounds are new, alarming in any way, or emotionally charged, the amygdala triggers your fight-or-flight, freeze-or-faint, or tend-and-befriend responses of the autonomic nervous system, and you might become defensive, smile nervously and over-talk, leave the room, shut down, or try to be helpful.[1] For example, when composer Igor Stravinsky premiered *The Rite of Spring* in Paris in 1913, the auditory and visual shock of the unusual sounds and dancing sent audiences fleeing. Now, many of us are familiar with atonal sounds and angular movement; it is no longer new.

Sound is central to the "feeling" of a culture and location. Complex reactions occur based on your past experience. This embodied "feeling" or relationship comes from sensory input, and sound is one of the significant components. If you've lived amid the lively voices of Italy, the quiet wide-open spaces of the western U.S., or the urban rush of traffic in Mumbai, those auditory "feelings" will be familiar. The sonic landscape has impact on communication: acoustic qualities of the room where you meet—including the hum of electricity or neon lights in a "smart" classroom and architectural components of the space itself—often dictate what happens. Some rooms are "dead"—absorbing sound and offering little

Silence

"Let's talk about silence," I say to a class of eighteen lively students. "What's your experience of not speaking?" One describes being raised with two sisters: "They talked all the time and I listened." Another said, "I like to formulate my thoughts fully before speaking." An Icelandic colleague from the communications department was in the classroom taking photos. She adds: "At home we were given permission to not engage. And when my father calls on the phone, it is often silent. We understand each other without many words." Another student admits: "Sustaining silence is hard. I'm taught to fill in any empty space with words and ideas. When someone is silent, I feel uncomfortable." For the final group projects, they practice listening—giving time for quieter voices to articulate their views. One young man surprises us all with a skillful monologue. What happens when you make space for others? When you value silence as a form of expression and connection?

Twenty Voices: Keep Going
Rehearsing my hour-long solo for an
endowed chair's lecture at Middlebury
College, voice coach Dana Yeaton is
tireless. At an early run-through, he says:
"You are giving me three Andreas, I want
twenty." One week before the presentation,
I fly home to Vermont from California in a
New England blizzard. After that cross-
country journey, the plane is rerouted from
Burlington to Hartford, Connecticut, so I
find a shuttle, then catch a bus traveling
north from Boston toward Vermont. Five
hours later, we arrive in Burlington in
whiteout conditions—you simply can't see.
My stalwart husband picks me up for a further forty-five-minute snow-bound drive
home. When I call Dana on my cell phone,
he says, "See you in an hour for rehearsal."
We work for two hours, expanding and
critiquing voice. There's nothing lax about
vocalizing and art-making; we go all out for
what's possible.

reverberation—while others resonate and amplify the warmth of vocal
tones.

Present since the origins of humans, music is a highly visceral partner
in embodied communication. The endocrine system is highly responsive to
sound vibrations, with specific glands responding to vibrations of different
pitches and frequencies, releasing a chemical bath affecting emotion and
mood. High-pitched sounds with short sound waves like a police whistle
resonate in the small sinuses of the skull and often stimulate alertness; slow
mid-range tones and melodies like lullabies resonate in the spacious heart
and chest region and can be calming; low tones with long sound waves like
bass drums vibrate in the spacious resonating chamber of the pelvis, invigorating the impulse to move your legs or at least tap your foot. Musicians
play the room and all the bodies in it. You are touched and manipulated
by the vibrations happening around you. Breaks, pauses, and dynamic
emphases in sound all have impact. And tone of voice affects response
even more than language: many children learn to distinguish when their
mother is really mad or just telling them to come inside based on tone.

LANGUAGE

There are challenges and benefits to using verbal language to transmit
what we feel. We use words to point toward or guide an experience, not
to replace the experience. When articulating what we feel about an event
or a person, we are translating sensations and feelings into words to communicate to others and to clarify those sensations for ourselves. Meaning
is transmitted as much by how you speak as what you say. Similarly, in the
process of teaching others, including facilitating an experience like the
process of relaxation, we use language to guide a state, mood, or pathway.
There's skill in matching experience and words—finding just the right word
and tone to convey or evoke.

Words have power. They can invite you into an experience or slam the
door shut in a conversation. They can also hurt and have lasting impact.
Words land and register in body tissues. In order to have empathy and
understand how our words touch and impact others, we have to feel how
others' words touch and impact our own bodies. Empathy involves the
expansion and constriction of our body boundaries—how close or far do
you want to be in relation to others in conversation or presentations? In
embodied speaking and listening, empathy is part of feeling what you are
saying: being inside rather than outside of your words, speaking from the
body, not about the body.

Exclusion has the opposite effect. Part of the skill of listening is noticing what isn't being said, who isn't included. Bringing the "silent voice"
of someone less comfortable speaking in a group into the weave of conversation helps make connections. Shaping conversation is a creative act;
many hosts of dinner parties know this well. The exchange is not overly
controlled but has intentionality and shape and encourages inclusivity.
Often a central theme emerges that can offer cohesion.

The skill of conversation involves picking up and following the thread of the person speaking and taking it somewhere new. At an embodied level, this includes sensing and feeling what is occurring in the bodies of the people involved, the active improvisational weave of physical and emotional exchange in conversation. This might be reflected in the elevated heart rates of enthusiasm, the strong emotional resonance from limbic arousal, or the gestural expressivity that underlies and interweaves with the thinking-verbal exchange. This makes an interconnected web of conversation, a continuity of consciousness. Something emerges that is whole in the collaborative process that you couldn't make alone.

TIME AND TIMING

It takes some of us longer to formulate thoughts. This may be affected by native language, learning styles, or personality types. How do we make time and space for everyone to have a voice? How do you notice when you are dominating the conversation? Sometimes interrupting others is from enthusiasm; other times, it's just poor listening skills. You may need strategies for when you get excited about what you are saying, personal rules about how many times to speak. Self-cues can remind you to engage whole-body listening: feel your feet and re-ground, jiggle your spine to release tension and restore fluidity, take a few deep breaths to refresh listening. Some Americans have grown up to be confident, taught to speak up even if they don't know what they are talking about! Others are self-criticizing; by the time they formulate their thoughts, the topic has passed.

Really listening well to someone takes clear intention. One simple technique to try involves softening the jaw. Jaw tissues pull on the inner ear and disrupt hearing; you can be more receptive with a relaxed jaw. People in sales often learn that a relaxed, slightly open jaw cues communication: you are available for hearing people's stories. Offering a receptive face rather than a closed, judgmental mask invites opening to the space between you. Air flows in and flows out without interruption. Lifting the eyebrows and giving a slight, authentic smile have been shown to stimulate responses in the body for alertness and connection.

Customs for speaking and listening reflect individual and cultural identity. There is no one right way to communicate in an embodied way. In one context, it might be important to speak up; in another, it might be better to restrain and listen more attentively and substantially. As you refresh embodied intelligence, your speaking and listening skills increase. You can become more conscious about connections, allowing circulation between what you hear and how you respond without becoming self-conscious.

Connections

Your lens today focuses on conscious listening and playful vocalization. A few times during your day, sing a song, recite a poem, hum, or tell a joke to feel yourself speaking and sounding. The key is to remain relaxed in your whole resonant body. You may need to find a private place indoors or outdoors to play with vocalizing—eliminating strain or restriction. Shake out tension, invite connection, and be gentle if you need support. Voice

Vocal Recording

We work with three speakers to record the voice-overs for our *Body and Earth* films. One is skilled equity actor and vocal teacher Alex Draper. I watch as he records my written text. First, he reads through the entirety carefully, asking questions to clarify. Then he shakes his full body to get rid of tension. When ready, he plants his feet firmly and signals the technician to begin recording. Making an error in intonation, he pauses and repeats the line knowing it will be edited. I watch, learn, practice for my turn at recording. I need multiple "takes" to find one that feels embodied. I require this process of students in our course on communication: record your voice and post it on your website. "I hate my voice," is a common response. Yet these folks are heading into public leadership positions. Moving past ego response to develop skill, you attend to rhythm, breath, and emphasis. Voice is potent; podcasts can be more powerful than videos because humans are so attuned to vocal innuendos. Recording voice involves practicing the roles of both speaker and listener.

Deep Listening

My husband is telling me about his meeting as we walk one morning. On one track I hear his ideas, on another I intuit that he should be home for a phone call, and on an emotional level I feel the strain in his voice. Circulating between these dimensions, I might just identify my appropriate response. Skillful communication—deep listening—is acknowledging all the levels and discerning your best reply, which might be to say nothing. You can listen to someone's words on many levels in the same conversation. The challenge is to not want to fix something for someone—just listen.

Just (Pause)

Speaking on a panel about "Embodied Voice" at a Body-Mind Centering conference, colleague Lisa Nelson in our prep time says: "I don't like panels. Nothing in depth ever happens." She proposes that we create a score where the audience can call out "pause" at any point in the conversation. The panelists can also call "pause" if we want to absorb or think about what is being said. When the pause feels complete, the speaker can continue what they are saying or shift to a new topic. (*Pause.*)

Lisa adds that a conversation is a contract between the speaker and the listener; there's a relationship implied. Panelist Bonnie Bainbridge Cohen reflects that she didn't speak for the first three years of her life: "Sometimes I feel like English is my second language, but I don't know what was the first." Lisa considers the possibilities in her inimitable way: "It's a long way from my brain to my tongue." As sometimes happens, the pre-panel conversations are more interesting than the event! (*Pause.*)

is a powerful medium, worthy of your attention, so you can feel and hear what you are saying. Vocal range and resilience involve allowing audible sounds through the narrow passageway of the neck. There is always possibility for more ease and discernment in that intimate transition from inner experience to outer expression.

TO DO

STAND AND DELIVER (DANA YEATON)
5 minutes

Public speaking is most effective when you have a relaxed and responsive voice with a sense that you are aware and care that we understand what you are saying. This is a practice exercise for speaking to others.

Standing with a text that you would like to share (to an imagined partner or group):

- Plant: Connect your feet in a stable position to the ground before speaking.
- Tune: Connect to your specific audience.
- Speak: Allow the voice to reflect your interest in the material you are sharing rather than focusing on yourself.

TO RECORD AND WRITE

HEARING YOUR VOICE
30 minutes

A certain level of objectivity is necessary to hear yourself without judging. Choose something short that you have written in these investigations, like the story of your name. Select a recording device (cell phone, computer, video, etc.). Record your voice reading; then, listen to your voice. Repeat this process several times, adjusting your pitch, tone, and dynamics until you feel satisfied with your recording. Write about your experience of hearing your voice.

Peter Schmitz and Nancy Stark Smith
RIFF Talks (2014), School for Contemporary Dance and Thought, Northampton, Massachusetts
Photograph by Peter Raper

Writing and Reading Aloud

Writing is a process of putting words on a surface. Whether it is connecting to a screen or a weathered journal, there is an active link between brain and fingers channeling the split-second decision-making that guides writing. The body maps the objects we carry and touch—like a book or computer—as the body, extending body-boundaries into space. In much the same way, we inhabit the page! Little marks on paper or screen can clarify our thoughts, make meaning in other peoples' minds, and leave traces for future generations.

As an ancient form of communicating, writing transmits and translates ideas and embodied experience into symbols that can be read or spoken. These markings can be signposts for future generations, a record of others' experiences within the global arena, or a catalyst for present-day thoughts. Books, manuscripts, journals, and speeches have their own lives. A few words taken out of context can derail a project or career. Just the right turn of phrase can make an impact. Cultural context informs language use; expectations around speaking and communicating are conveyed by social and political norms.

Informal writing in a journal can be like taking a long walk with yourself. Similarly, some children and adults enjoy books that take them away from the life they are experiencing to imagine other ways of being. But the process of revision takes committed time and discernment. Often you have

Carnality sits at the root of the show-don't-tell edict that every writing teacher harps on all the time, because it works. By carnal, I mean, Can you apprehend it through the five senses? — Mary Karr, from "Sacred Carnality"

Kate Trammell in *Namely Muscles* by Claire Porter
Photograph by Richard Finkelstein

Journals: All Those Words

I am destroying forty years of journals, the same spiral-bound beige notebooks featuring pale green, narrow-ruled paper. Piled in closets, on shelves, and in boxes under beds, they fill every niche and cranny of our small Vermont house. When someone in my women's writing group asks what we'd like done with our journals after we die, I know mine are never to be read. (I put in my journals what I don't want to be thinking!) So my husband and I begin the process of disposal. The wood stove is too slow; recycling offers risk of retrieval. When the giant dumpster is delivered for our move to Maine, it's "heave ho." Journals fly through the air, merging with other discarded memories. At our even smaller nest in Maine, there are more! Now, I have a plan: every ten years they must go.

to let go of something to receive something new. This can mean giving up something in your day so there's space for writing or editing out a favorite word that appears on every page. My colleague John Elder describes the process of letting go of the unessential as *relinquishment*—to surrender certain elements in the interest of achieving greater clarity overall. It's a part of the writerly process that also applies to life's journey.

Regardless of your relationship to writing, you *will* be writing. This could be through emails, texts, blogs, tweets, or some other new form of cyber communication. We use words to clarify thinking, making notes for what we might say in a meeting or reflecting on what others said after the fact. Sometimes words come as if we are in a trance—they tell you what to say as if you are a channel, a tool in the process. More formally constructing your views involves addressing your writing more consciously. Considering how, where, when, why, and for whom is part of communication strategy. The degree to which your whole body—whole person—is involved affects your authorship and authenticity in the written medium.

THE WRITING PROCESS

Writing is a practice, just like any other practice. Skilled writers develop a healthy and discerning inner witness, judiciously deciding what stays in a work and what gets deleted. The discipline requires balancing your more self-critical, workaholic, and perfectionistic inner demon with your play-it-safe and tone-it-down best friend of an inner voice. Every author who invests in personal stories confronts the specter of ego partnered with outer criticism and inner insecurity. Mary Karr in her book *The Art of Memoir* writes, "All those vertical 'I's' can dot a paragraph like slammed doors." How to be personal and compelling without "the ego dragon crashing through the room?"[1] Cultivating a relationship with your inner witness as a discerning friend can support a healthy sense of proportion.

Cultivating an external witness, like a good editor, teacher, therapist, or a disciplined writing group, is one step on your journey. In some cultural contexts historically, you had a guru (teacher) or spiritual guide who kept you on track or a shamanic mentor if you were taking journeys to "other worlds." As writers, we often use our friends or family in this role to support our inquiry, but it's not really their job. Working with an appropriate outer witness offers feedback without damaging your authentic voice or fatiguing relationships.

Your participation in this disorienting process we call writing is dynamic, and change and grief are active themes during the timespan of a written work. Amidst all the beauty of everyday life, children move away and shape their own dramas, marriages transform or crumble, dear friends die, the world remains at war, and the environment is under constant attack. Challenges to mind, body, and heart occur—these things we humans do to ourselves and to others. The difference between psychosis and the artist's and mystic's journey may involve strong boundaries and a healthy container. Exploring unconscious, repressed, hidden caves and passageways in ourselves requires a clear tether back to daily life and a practiced form, like writing, to communicate our findings to others.

There is a dialogue between specificity and evocation. Details and close observation matter, as do imagination and transmission. Captivating facts and details lead us to believe in a writer as well as to undertake our own journey of discovery. As readers, we need to inhabit our personal views and sensual embodied selves. Author Barry Lopez, in a writing workshop, stated that in the beginning of a novel, the author has their hand on the reader's back as guide; but by the end, it's the reader's journey and the author has disappeared.[2] To do this, we need enough information to build trust without creating an overwhelming shopping cart full of details.

SKILLS

In writing and reading aloud, you can change temporal and spatial scale to invite multiple perspectives. It's like working with a camera: you can zoom into detailed description and then zoom out to a larger perspective. You can focus on the tasty choice of the words themselves or language as a cultural lens engaging a meta view. Etymology reveals context: how did a word come into being, and how is it interpreted in current usage? Part of a word's value is that it slows you down, focuses your busy mind. A good word, like "capacious"—referring to a topic in writing that can take everything you throw at it—can make you pause, taste, digest, wonder. And then there are phrases like "serendipitous delight" or "spontaneous joy." You feel them in your body—a sensorial response.

The sonorous dimensions of language are often forgotten until you arrive in a new culture with a new language that initially registers for you more as sound than meaning. You do not know what someone is saying, but you hear it, feel it, and guess. There is no direct route to meaning through the language part of the brain, so other regions take over. The power of a word or phrase includes its sound, its sonic impact, its context. How can the written word replicate or imply tone of voice? Writing is often accompanied by design elements or other expressive mediums to add color and auditory expressivity to the written page: photographs, white space, typeface, and video imbue the experience of the words and the information with dynamics. In this way, we broaden the sensorial possibility of words to touch the reader, embed feeling in the imagination, and move meaning into action.

MEMOIR

Writing about yourself has its own parameters. With a focus on personal narrative, you are excavating your stories and others' lives as well. The genre of memoir has inherent opportunities: it offers a sense of freedom and authority—it's your story, your chosen timeframe! And it has challenges: some writers get lost in the self-analytical process during the years it takes to accomplish a project. Yet personal narrative can also heal; narrative medicine uses storytelling as a way to move damaging stories onto the page where you can become more conscious of them and move on. Our

Writing Places

To finish anecdotes for my *Body and Earth* book, I return to my father's ancestral homeland: Samsø, Denmark. I need a location that holds meaning and animates sensual delight. Staying at Pensione Verona, on the north end of this wind-powered island, I have one week to complete thirty-one chapters of anecdotes. At breakfast, I stuff an extra boiled egg plus cheese and bread into my backpack and bike to an isolated beach. As I alternate between writing, dancing, dining, and resting, only an occasional boat witnesses the process. In the late afternoon I change locations to the windy headland, where larks swirl overhead and vistas expand. Lying on the floor of my tiny bedroom at night, the day's scribbled writings in hand, I edit. As Maine author Edmund Ware Smith describes it, "working . . . on the pencil-pecking for my book."[7] At some point the question arises: does anyone need to read these stories? Of all the voices that emerge on my island sojourn, that was one to ignore.

Notepads

One thing about stories: you take what comes. There's no waiting for a more convenient time. Author Julia Alvarez encourages young writers to always carry a notepad—a pocket-sized receptacle to record specifics of a moment. Terry Tempest Williams in her book *Leap* describes jotting down revelations about the medieval triptych painting *The Garden of Earthly Delights* by Hieronymus Bosch at the Prado Museum in Madrid for six years. Musician Paul Simon, in a radio interview, reflects that if a song comes into his mind when he's driving, he pulls over immediately to write it down. If he waits, it's gone.

The Night Before

All writers make self-rules—which we stick to or not. John Elder, whose writing practice starts at 4 a.m., states: "I don't read emails before breakfast." Poet Robert Pack, former director of the Bread Loaf Writers' Conference in Vermont, is careful about evenings. He listens to music rather than socializing. Poet David Whyte poses challenging prompts: "What's the question you are most afraid to ask?" For years I would read Barry Lopez's incisive words before sleep so I would write with more dash and precision the next morning. Sometimes it works!

Finding Your Writing Voice

A student from the Gambia was taught that writing involved repeating *precisely* what he had read by others. Confused by the thought of writing from his own voice, he was unable to complete our first assignment, his personal "placestory" in our writing class. Weeks went by past the due date. Recognizing the sincerity of his impasse, I encouraged him to imagine he was writing a letter to someone he loved— his mother or a friend, not a teacher. He wrote fifteen eloquent pages for a seven-page assignment, articulate and clear. He taught me to never take freedom of speech for granted; it is hard won and needs to be scrupulously protected.

mental maps impact our health. Doctor and philosopher Joan Borysenko writes: "The stories we habitually tell ourselves change our brain circuitry, our relationships, and our path through the world. They can elevate life to heaven on earth or drop us unceremoniously into the fire."[3]

Who owns a story? Especially when working in countries and cultures unfamiliar to your own background, ownership of stories is important. Who has the right to tell a story? Every individual in a family who has witnessed an event has a different version and is likely wedded to specific views. Memory is distinct but not necessarily accurate. Bits are interwoven in tissues (like the fascia, which holds postural habits) and stored in specific parts of the brain (like the cerebellum, which retains movement memories), and these can change. Perception is a construct. It is filtered through your past experiences and embedded in your personal neurological network. That's important to know in communication. When using others' stories in your writing, send them out for confirmation. Maybe one or two words will be tweaked; seldom are whole stories deleted, but if they are best left unpublished, it is worth knowing before print.

EXPLORATORY WRITING

Journaling is often a base from which writing is drawn. The personal process of exploratory writing moves beyond lists and precise records of what happened when and where (although that can be essential) to making connections between various life experiences and perspectives. Described by author Ken Macrorie as "writing freely with a purpose," exploratory "free-writes" in your journal involve writing to discover, strengthening your capacity both to understand your own values and goals and to communicate them to others.[4] When done with fellow explorers, this process lets you consider a writing group as a community of learners cultivating a community of writers. Feedback in this context is not about evaluation or praise. It probes, "where can this go?" and creates shifts in the possibilities of your writing.

Some of us write to hear what we are thinking; to discover what wants to be said. The encouragement from a writing coach, like Natalie Goldberg in her popular book *Writing Down the Bones*, is to put pen to page and simply begin—no need to overplan. Find a café or back bedroom, write for ten minutes, and don't stop. Then look at your last sentence, rewrite it on the next page, and start again. The stories that want to emerge will come through regardless of the immediate assignment if you don't get in their way. You don't have to "start." Thoughts are already flowing. We talk all the time. You can utilize some of the velocity of speech to propel your writing voice. Writing rapidly, energetically, and associatively, sometimes you hit a rich vein that lets you reach deep, find something galvanizing— even astonishing.[5]

Personal stories and anecdotes lead us to the body; there's a visceral and emotional response. Stories capture the part of the brain that wants to know what happens next, keeping you turning the page. Essentially, all good fiction asks one question, "who changes and why?"[6] As a project moves out of the head and into the heart and body, it becomes empathetically impactful. Reading your writing aloud can enhance that process, as can the inclusion of resonant stories that move the personal to the communal, inspiring emotional connection through the felt landscape of the voice. Revision—the essence of the writer's craft—helps you weed out repetitive or uninhabited language.

Referencing other people's work can give you a sense of authority: you are well-read in your field and have (hopefully) assimilated their views. It can also cue insecurity, as though you need to prove what you are saying by justifying your writing with snippets of others' works. It helps to sequence references consciously; consider international, intercultural views; and have a sense of chronology, location, and context. When did someone write and where? What external conditions and discoveries motivated their views? A twelfth-century mystic in Turkey and a contemporary author in California have different contexts even though some of their universal content may overlap. It's compelling to show solidarity and interconnectedness with other writers and to discern why you are bringing in their voices. Quoting other views can broaden the reader's capacity to hold more than one perspective at a time.

SOMATIC WRITING

How do we inspire people to write from their embodied experience? Each new generation of writers takes us into the future. Words carry the culture. How can we create a field of somatic writers so this view can go beyond the present day? Historically in most Euro-American cultures, the voices of men were prioritized above those of women, and we see this in the way literature is taught in schools and the proportion of ancient writing attributed to men versus to women. Now, with more women authors and diverse voices being prioritized, there is a different kind of writing—we are just touching the beginning. Of course, there is a lineage of women's voices in print; some wrote and their words got lost or were excluded. What kinds of literature are we leaving for the future?

Embodied writing can be a stepping-stone to get you into or out of an experience. It is part of our lives in multiple ways. Watching a child learn the process of connecting thoughts on a page can seem like magic, and it has some of the same implications. As we recognize the power of words to heal, to hurt, to shape history, we can notice how we utilize this medium in our communication skills. Embodied writing means that we feel what we say and recognize the impact words can have on others, including attitudes about other peoples and species.

Connections

Your focus for today is on writing. This may bring a feeling of pleasure or resistance. Set yourself up with pen and page in a suitable place and allow yourself to write *from* your breathing, sensing body, not just *about* your body. What time of day invites your most visceral writing voice? Rather than assuming the answer, give yourself the opportunity to write several

Revising

Poet Syd Lea offers: "Curiously, it is by way of revision that I discover what engendered my poem; but the pleasure of revising can be addictive, and there comes a point when one is indulging it, not to improve the work, but for the pleasure itself—and one must simply stop."

Reading: Action/Phase 1

June 13, 2020: Choreographer Bill T. Jones reads outdoors at New York Live Arts during the Covid-19 pandemic and racial protests, "In my isolation, I have gone through my library, which is in some way revisiting a pantheon of heroes, villains, the mausoleum of ideas, and the maternity wards of notions of what the future might be. I am reading as a way of thinking."[8]

Writerly Views

"Language doesn't just reflect the world, it creates it," writes Eve Sedgwick in *Touching Feeling*. John Updike in an interview with Terry Gross says: "A writer is someone who tries to tell the truth." Most authors agree with English author Samuel Johnson: "If you want to be a writer, then write. Write every day!"[9]

Tula Isabel, morning writing
Photograph by Karen Clarkson

times today. Early morning is close to dreams and less cluttered with the events of your day; midday you may be alert and fresh; for some, evening can lead to a calm quietude or to more passion and expressivity. Everyone's life is different. If you get stuck, that is part of the process. You are practicing resilience. The invitation is writing to discover, exploring your expressive range through the written word.

TO DO

JOURNALING: WRITING TO DISCOVER (JOHN ELDER)
30 minutes

Taking yourself somewhere new can increase concentration.

Go to a café where you can sit for a long period without interruption. Write about your morning. Describe it in detail, zoom into specifics—the wake-up alarm, the smell of coffee, the restlessness of deciding. Look for some edge zones or cracks that take you into your passion place. If you get stuck, begin writing: "When I write I . . ." "When I write I don't . . ."

TO WRITE

LOOKING BACK
30 minutes

You've been writing all along. Now, reflect on the journey. Read back through your writing and notice any emerging themes or repeated questions. Continue writing from that source; take it to the next level. Let the writing format emerge: is it an anecdote, blog, screenplay, fiction, non-fiction? When you are done, read aloud and notice your writer's voice and the framework that emerges. Reading helps with editing; you feel what is essential and what can go away!

Embodying Digital Media and Filming

With Scotty Hardwig, digital media artist

Digital media affects how we think as well as how we interact. Communication is increasingly online. Screens link doctors to patients, teachers to children, and family members to each other. Birthdays, memorial services, and festivals—everything, it seems, but haircuts and dentist appointments—are now virtually possible. As the digital becomes daily, conscious awareness of the body's role in communication is increasingly important. Digital engagement questions our relationships: how much do we give over, and how much do we maintain agency?

The ways we present and represent ourselves in virtual space create a digital identity. Often, we have multiple identities online according to which space, platform, or group we are engaging. The ways that our bodies and the bodies of others are depicted on screen affect how they are perceived. Aspects of embodied presence always register—from fatigue to attitudes and posture. If we are framed in a virtual box, filmed outdoors, or

If you are purposefully putting yourself on the web, what do you want people to know about you? — Gardner Campbell, from "Romantic Computing"

Time Garden: Noon (2020)
Virtual reality choreographic work
by Scotty Hardwig and Zach Duer

Body, Full of Time (2018–2019)
Motion capture, projection,
and interactive avatar designs
Image by Scotty Hardwig

Communicating Now

I'm confused about communication. Cleaning house I begin discarding cardboard boxes of files. Hesitating for just a moment before tossing, a folder of letters catches my eye. In it I find long, intimate writings by my father offering advice (and hesitation) about my becoming an artist as profession, a sister explaining her divorce, a lover conveying distance. These were missives to save and cherish. Mail now is mostly bills, phone calls are less frequent, emails interruptive, texts too short. And the plethora of options makes me pause: do I use Twitter, FaceTime, Skype, or Facebook—formats that constantly change and require computer updates and scrutiny? The impulse to connect is deep, but knowing how to sustain meaningful conversation in this digital era with weary eyes from too much screen time is problematic. My colleague who is an author says she writes about what she doesn't understand. As I look for words around embodied communication, I am pondering: How best to connect?

documented in a public setting, imagination and the capacity for empathy are impacted. As we choose how we want to engage with virtual media, we can notice their impact on our bodies, our self-awareness, and whether they enhance our well-being or distract.[1]

Embodiment can be facilitated by digital communication or diminished. We have access to a global community, and many resources for engaging experiential practices are available, from remote summits and learning ancient movement practices to focusing a digital camera and choosing a recipe for an evening meal. Many web-gatherings are experiential—the viewer is an active participant through chat functions and visual presence. Yet situations where it seems like we are invisible and removed from physical engagement are often deceptive. Like in life, even when we think we are in privacy, we are often being seen by someone. Information we enter into websites or our smartphones becomes part of data collection that can be used to shape our choices and focus our habits. Keeping control of our data means being aware of our interactions. Where do we live online? Do we visit a site often or just happen upon it? For many of us, more people are affected by our online presence and virtual identities than by our physical interactions and conversations in a day.

We are not separate: the brain maps objects around the body as the body—your hair, clothing, and the cell phone consistently in your hand are part of you.[2] Fingers on your computer keyboard extend your body map; this extension of your embodied self can feel useful when writing, viscerally connecting you to your words. Hearing cues the nervous system to the vibration, ring, or ping of messages arriving, affecting focus. In fact, digital scholar Amy Collier describes contemporary humans as "hybrid"—with digital devices as extensions of the body.[3] We are human plus . . . There is a particular impact on our brain and nervous system, and embodied communication is affected. Research on health implications has yet to catch up with the speed of increased usage.

HISTORY AND NOW

Since its inception, the internet has offered a virtual Wild West of human communication. This includes never-before-seen levels of anonymity, disembodiment, and imagistic consumption in human communication and commerce. From our beginnings as a species, our *techne*—our artistry and techniques for making and expressing relationships to our environment—have become our technologies. This may include tools for carving and harvesting the raw elements around us, architecture for shelter from or worship of the natural world, and paths and roads for moving from one place to another on our journeys. The industrial revolution, with its towering machinery and combustion engines, vastly amplified the power of the human body to complete tasks that our muscles and sinews could not—manipulating our surroundings. The advent of computing, similarly, has vastly amplified the power of the human brain; computers are capable of completing tasks that our organic brains and nervous systems cannot manage. In this way, we can understand the invention of computing as a major revolution in human history; the effects are difficult to discern, happen faster, and are more impactful. One way of describing the digital revolution is doing very old things *faster*.

The internet as a global network can also be seen as a new extension of the age-old, primal impulse to meet, make, and connect. More than just an inherently disembodied phenomenon, it can be viewed, reframed, or experienced as a deeply embodied stream of connectivity. Through technology (specifically the language of computer code) we can move through time and space, make a new and different kind of reality, and experience reality in a different way. Are we ready? What are the questions being raised in those spaces? How does the digital realm limit or support communication?

There can be struggle involved with the desire to see beyond what we are able to inhabit and integrate. Depression, clinically, can come from feeling that there is something very real that is out of balance with this world. Alienation can occur when we are not physically present together, when the energies and vibrations of our bodies are distant from one another. How do we question and still stay connected? We are all involved in some way in these inquiries every time we open the computer, click on a message, or purchase something on the internet instead of at our local store. How do we live with full embodiment and engagement within our home communities (people and place) in the age of cyberspace? A sense of humor, a bit of irreverence, and recognition of the absurdity of too-muchness can be useful antidotes to overwhelm.

The sharing of work on the web is meant by definition to be open to people around the world. In the vast and interconnected network of digital thought-connections, each individual has the potential to communicate and share ideas with all of the others. Information and the transfer of information is carried out from each-to-the-other rather than from the top down or from a center reaching out to others. This interconnectivity is in many ways similar to interneurons within our nervous system and directly mirrors the way that Earth systems function. Meteorological activities, tectonic geology, and the water cycle offer models of interdependence, with one part relating to and impacting all other components within the system. Being critical of how digital templates limit our multidimensionality and interconnectivity is increasingly important. The digital medium is not fixed; it is made to change and adapt. But all those prescribed formats, forms, and menus can limit creative imagination, multidimensionality, and embodied communication. Understanding and redefining use of the tools is part of our challenge.

Mediated imagery is culturally coded. Advertisements on broadcast-based media give us a lens into the economically driven imagination of industrialized modernity. Images tell a story of what we are trained to believe we want, and we are invited to place ourselves within a cultural and aesthetic view of the body that may not reflect our own values. Selling desire-based rather than need-based products is not just an industry; it is a science and a seductive ploy.[4] Neuroeconomics is a field that studies and targets the very cells that create desire and addiction for commercial

CAMERA TERMINOLOGY
Scotty Hardwig

Exposure: general term for how much light is let into the camera lens, resulting in how dark or bright the image is in a photograph or video.

Aperture: opening through which light is let into the camera lens. If the aperture is bigger, it lets in a lot of light; if it's small, less light is allowed in.

Shutter speed: measurement of the speed of the mechanical shutter in front of the camera's aperture as it snaps open to allow light in momentarily.

Framerate: measurement of how many individual frames will appear per second in video recording. Videos are essentially a series of still photographs set into a sequence.

ISO: general term for the sensitivity of the camera lens to light (or traditionally, a measurement of the chemical mixture on film strips for how fast it will absorb light).

Color temperature: setting to tell the camera how to identify the ambient color temperature of the environment during a film shoot.

Focal length and depth of field: measurement of which parts of the shot are in focus and which are not. This can be adjusted using both the zoom settings and focus rings on the camera.

Scotty Hardwig (2018)
Monterey, California

Just Be Natural

My first film is a student production. The takes and setups include appearing to steal food at a local grocery store (Greg's Meat Market), hiding in a dress shop (Wild Mountain Thyme), and disappearing with a shopping cart in a night-time parking lot (outside the Middlebury Co-op). "Don't act, just be natural," the director advises. "Let the camera do the work." Okay, but that is easier said than done with a five-person film crew, lights orienting focus, and multiple repeat takes for different angles. When the film is screened to a packed audience, the credits list my character as: the "Old Lady." Even with cheering fans, that surprise squelches any shred of ego!

purposes. Advertising employs a host of graphic designers, videographers, editors, color graders, photo editors, lighting designers, models, directors, dramaturges, actors, and even dancers to lure our attention. In these fantasy worlds of image and message, the relationship between subject, object, and viewer is carefully chosen and curated—it is mediated.[5] The ethics of the image (what we believe to be right and good or wrong) becomes collated with the aesthetics of the image (what we believe to be beautiful or ugly), and all of this becomes the engine that drives our current economy and communication norms.

But the forces of mass media and mass persuasion through imagery can also be harnessed for agendas other than monetary ones. Digital technologies can be a vital part of environmental or social activism. Social network analytic tools can be utilized to uncover patterns in communication and relationships. Increasingly, every person has a camera. When portable devices and smartphones allow individuals to instantaneously take videos and share moments online, we have the ability to see subjective experiences of individuals in the moment, whether they are filming landscapes, friends, aesthetic events, or images of police brutality and social injustice. With such an immediate connection between the individual eye and the public eye, what social changes might evolve in this new reality? What will we see differently?

How can the digital eye of the camera, our presence as videographer, or the relationship between web-based presence and live activity deepen our understanding of a person or place? Can filming in natural sites that are hard to access in our daily lives offer opportunities for connection with the elements? By working with somatic, embodied experience through the lens of the camera, we can begin to understand multifaceted ways that our digital worlds can once again be the link between our bodies and our Earth. We can seek an embodied permeation of body and landscape, flesh and data, transcending the barriers between skin and nature to inspire respectful inhabitation, feeling ourselves and the world around us.

FILMING AND BEING FILMED

Filming is one way to build empathy and fluency in embodied digital media. The camera is an extension of the human eye that replicates how the eye sees. When filming something, you decide how to frame a viewpoint. (How do I want to see this event? How do I want viewers to witness this event?) A particular theoretical understanding of the dynamics of power is implicated in the choices made with a camera. There are multiple histories of the same moment. You have to select your perspective, map it out in space and time, and be willing to listen to collaborators. Most videographers prepare a shot map; there is an overview in mind and an ethical framework. Improvisational adjustments happen in the moments of filming to respond to conditions. An openness toward experimentality and improvisation is critical in developing an aesthetic and identifying your point of view. Deciding how your viewers are going to experience this event is radically powerful; there is a decision-making process involved.

Understanding the craft of filming involves technical knowledge of how the camera works.[6] It takes time to become familiar with exposure, focus, frame rate, and all the specifics of the equipment you are using—whether phone or camera and tripod or gimbal system. The craft of the technology is part of the play of discovery. For example, the way you set the camera to absorb light determines the outcome; the mechanisms of the technology and the product are interwoven. Where technology becomes poetry is in being familiar enough with the mechanics that you can focus on selecting your viewpoint. Questions are inherent: how shall we go together on this journey, this experience of witnessing? Selecting involves balancing your subject's interests with what your interests are as an artist and with what an audience's interests might be. Editing the hours of raw film footage is the place where the flow, pacing, and creative perspective of the scene become fleshed out. The rhythm of the edit is different from the rhythm of the shoot.

When choosing a film site, you can look at the energetic design of a space in terms of invisible forces. Generally, a videographer looks for spaces that have the least amount of noise. Noise is the amount of geometric or energetic clutter in a space, particularly as it relates to compositional framing. You can notice how each element within a spatial frame interrelates and the flow of energetic forces between them. Filming at a forested site in Wales, for example, this could include the elemental geometry of the space: how things are framed in arcs, lines, circles, and spirals. Simultaneously, the effect of gravity on vine-draped branches, the wind moving the leaves, and the plant life spiraling up toward the sun inform the composition. Filming can reveal how this space has grown into being, offering a feeling for its history as well as what remains. Embodying the craft of the camera in urban landscapes, you add the ways space has been constructed by human architectural design. Markings, graffiti, tracings of human use tell their own story just as bodies can reveal a life lived, a history recorded.

The sonic landscape also has impact. Sound conveys the size of a room or site: you can tell if the event is in a hallway or grandiose church by sounds of footsteps or voices without even needing the visual scale. During filming, you record the "sync-sound" (synchronous sound) of the room; each space is distinct. And in editing, there is a choice between using sync sound (sound captured directly) and later additions: recorded text, background effects, and musical compositions. The energetic flow of landscape and inhabitant body is inherently affected by sound. We can combine these elements to include all the sonic layers: the original sounds of the event, the dramatic effects and musical score, and the pedagogical or informative material.

Including other people in the frame of your images involves getting permission. The eye of the camera can be revealing and invasive as well as affirming and supportive of communication goals. Whose image is this? Who has the right to tell this story? When working with images, you need to credit the artist—who took them and when? Are they copyrighted or in some way given with permission? Recognizing the power of images to

Avatars

I didn't know about *avatars*. The possibility of two-dimensional and three-dimensional graphical representations of myself as a user in computing was new to me. Then I attended a talk by digital media artist Scotty Hardwig on this topic. He described the importance in his background of creating an electronic image, manipulated by a computer, that embodies a separate self on screen. As he was growing up in rural Appalachia, it expanded both who he was and could be and the worlds he could inhabit. The web became a source of expanded identity, of making connections. Having this alternate persona (or several of them) was freeing.

Screen Time

In our digital media series *Body and Earth: Seven Web-Based Somatic Excursions*, our goal is active and international—as far as our budget can stretch. Working with artists from seven countries, multiple languages, at four sites, filmmaker Scotty Hardwig takes lengthy shots, giving the artist time to drop into the specifics of place. At Marloes Sands Beach in southwestern Wales as we sit eating lunch, dancer Susanna scurries up a rock face; Eeva-Maria moves into the waves. Camera in hand, still chewing lunch, Scotty is off to the sites. Careful not to include passersby who might object to being filmed, he moves with and around his subjects—a steady cam smoothing out transitions. Dropping into a movement with a close witness is part of improvisational training, letting the site move you in ways you can't imagine.

Editing, of course, is another matter. Filmmaker Erika Randall says there is one film you plan, one you shoot, and another you edit—they aren't the same. That's true with a manuscript as well. Once you hand it to designers and final editors, the book develops its own life and feel. It's a bit like sending your child to school; other influences prevail. It's all part of the process of creating, from imagination to actualization; from film script to screen.

our last aria (2017–2018), duet dance
for camera by Scotty Hardwig
Performers: Keanu Forrest Brady and
James Mario Bowen
Eureka Dunes, Death Valley, California
Still image from videography by Michael Ryba

shape our view, we can be more aware of our choices. Filming and taking photographic images sometimes replace our commitment to presence: "I'll take a picture now and think about it later." In a lecture about digital communication, Gardner Campbell asked: "Would you want to visit this place if you couldn't take a camera with you?"[7]

When filming movement and embodiment practices, there is hope that the digitization of the moving or dancing body or the physical practice itself can be seen as a link that connects us to each other and the environment. By filming movement and being filmed, we try to find the link between technology and the human spirit. We cannot claim to fully understand the subtleties of these connections, but mirror neurons, kinesthetic resonance, and transference are all involved. Films can be education tools as well as art-objects, placed within the multiple inroads of the web. For those who wish to extend their body practices and better understand their relationship to the natural world, framing with a camera can be a pathway to empathy. We can dive into the processes of filming and being filmed as experiences of artistic integration linking digital media and embodiment.

Connections

Refreshing your relationship to your digital identities can take some imagination. Today the invitation is to take a selfie at various times during your day, connecting what you feel to what you look like visually and where you are physically. Expand your range in how you frame the images: try moving off center or focusing on background and context rather than on your figure as foreground. Then put your cell phone out of reach for a few hours without touching or checking. Notice your connection to this embodied living and communicating through the digital realm. Resilience in the digital age includes knowing when to engage and when to take breaks. What are your digital meals for the day, and how do you digest them?

TO DO

FILMING FROM CENTER 👁

30 minutes

> *Rather than just reading about creating photographic and videographic content, become a "media maker."*

Make a short video on one of your digital devices. Choose a subject with movement and physical action: join your center (belly) with what you are filming so you are moving empathetically and kinesthetically with what you are viewing even if the camera is still. Experiment with time, distance, perspective, and sound. Play with any or all of the "Five Central Elements for Media Makers": framing, exposure, depth of field, movement, and editing. What do you find?[8]

TO WRITE

UNPACKING THE MEDIATED EXPERIENCE (SCOTTY HARDWIG)

15 minutes

> *For many of us, digital involvement shapes our days. How can we be more intentional about influence and impact?* Look at a digital image from the web, an advertisement, or YouTube video. Spend some time with it. Describe it from the following three perspectives in the following order. It can help to separate them out into three separate columns in your notes to distinguish between different levels of conscious interaction. (10 minutes.)

1. *I observe.* This includes a detailed description of everything you see, hear, taste, touch, or smell as objectively as possible (without judgments or even analysis—those come later). The more specific and precise the details, the deeper your perspective becomes.
2. *I think.* Based on what you know, write about what this content might mean on an intellectual level. This can include its sociocultural values, connotations, and hidden contexts or intended messages. Remember, all human-created content has a foundational value system, set of shared constructs, and an agenda of some kind! It may bring you back to the thoughts or words of a mentor, scholar, or writer or a theoretical framework that you've read about in the past. In this column, historical context matters.
3. *I feel.* Emotional responses, "gut reactions," and value judgments go here. If you have an intense emotional reaction to something (either positive or negative), it can be useful to ask: "Which of my core values and needs are being met or not being met by

See Me!

The little boy cries, "Look at me! Look at me!" He's daring to enter the waves, his arms flapping up and down in a mixture of joy and terror. His mother is nearby on her cell phone. Determined to get someone's attention, he turns to me, a passerby on the beach. We share an enthusiastic grin—acknowledgement! That's all it takes to affirm his bravery; he turns back to the ocean, chin up, and wades deeper.

this experience?" We now have the benefit of referring back to the previous two columns for answers to that question. Often, we can be tempted to jump straight into this column after having an experience, but it can be useful to leave it for last, after undergoing the detailed work of the previous two. This way, we can learn more about our feelings and experiences, adding dimension and depth to our reactions, naming them, and exploring them.

Write about the process. Unpacking your response to imagery and advertisements can free impulsive reactivity and addictive behaviors (e.g., I see a hamburger ad, I go buy one). It can also enrich your experience of *every* experience by adding detail, depth, and discernment. Read your writing aloud, listening to how you have articulated your observations, thoughts, and feelings. (5 minutes.)

DIGITAL PROFILE: WHO AM I ONLINE?

With a partner: Share your website, Facebook, LinkedIn, or other digital identities you have created. Let your partner respond to the experience of viewing your digital profile: meeting you "online" and describing what they perceive.

Art-Making and Imagining

Day 15

Every innovative project is a creative project. You are making something that has never been done before. And if you're not changed in the process, the project isn't complete. For a moment, consider yourself an artist: not the ego of art display and performance but the process of filtering life through a creative lens and engaging the far reaches of your imagination. The ability to envision a project and carry it through to conclusion is central to artistic creation. You have an idea and manifest it while remaining open to unforeseen dimensions.

Through artistic training you practice making an idea or creative impulse into something real. You know that the process is likely to be multilayered and nonlinear and will involve significant challenges and ambiguity. Flexibility and tenacity help move you through the multiple layers of shaping, showing, and feedback without compromising intent. Ambitious projects require both resilience and a clear stance and point of view. As choreographer (and MacArthur "genius") Bill T. Jones articulates in an interview about art-making, you have to remain "voraciously interested in how to push the envelope forward."[1]

Arts modalities underlie daily communication; they begin in childhood with movement, sound, and visual marks and extend to more complex abstractions like language. By engaging different learning styles and

Being responsible for creating the future, it is important that we keep alive our capacity to imagine what that might be lest we continue to recreate the present. — Gordon Thorne, from a notebook

Every One (2012)
by Philip Buller
Oil on linen on panel,
60 in. × 80 in.

99

Museum Heritage

Art museums were my second home—heart spaces, fundamentally familiar. Every year growing up I visited the Chicago Art Institute with my father. He had been a student there; in our family, the quiz was to identify the artists and name our favorites. With this background, I was a bit shocked to be the first student ever in my study-abroad program in Paris to receive an F in Art History. It was during my junior year of college, and I spent almost every afternoon at the Musée d'Art Moderne. I knew the guards, the paintings, and painters by name, but writing my final paper "en Français" in my youthful enthusiasm revealed my lack of descriptive vocabulary and unfortunate disregard for grammar. The grade was revised (to a C, passable), but that evaluation gave me pause, thinking about the distance between loving art and the separate skill of describing why.

In the Artist's Words

Philip Buller, whose painting introduces this chapter, writes: "When I am engaged in the painting process, I am of course conscious of the sensations, thoughts, and associations. But rather than following them into the quagmire of willful, even enthusiastic intention toward constructing a certain meaning, I try to instead wait to be touched in a certain way; my feelings have to be stirred. The evolution of the paintings themselves seems to resist conscious intention. I try to remember that the meaning, if there is any, will emerge in its own time, leaving me with an awareness that the work cannot be hurried. When the painting is resolved, I feel as though something has arrived, like a fruit in an orchard, visible as an organic thing."[4]

honoring cultural values, the arts provide significant impact on the visibility and effectiveness of social and environmental projects, including modeling creative approaches to problem-solving and collaboration. Thus, they expand avenues for dialogue in local and international settings. Songs and melodies can touch our hearts even if we don't understand the words.

Often the arts reach for something beyond conscious awareness. They seek to evoke universal dimensions. In community projects, they provide a focus that is non-sectarian and interactive, a place to meet beyond controversy. There is a dialogue between the unconscious and conscious mind and between ordinary and extraordinary realities across time and place. For example, the art displayed in "Te Māori," an exhibit at the Metropolitan Museum of Art in New York in 1984, included distinctive carvings, paintings, and woven pieces that were honored with chants, prayers, and presence by Māori elders at dawn before the exhibit could be opened to the public.[2]

HISTORY

The human urge to create communicative symbols and leave traces has existed throughout recorded history. This impulse extends beyond functional tool making and speech to manifest as sculpture and visual images, sounds and songs, movements and dances, poetic or narrative stories and myths, masks and ritualized adornments. Aesthetically powerful images are not just about reporting and documenting, although those dimensions are important. They convey an unspoken message larger than the sum of its parts—moving beyond representation to transmission.

The arts are part of cultural cohesion. They are woven into a web that holds things—people, places, communities—together. To destroy a functioning culture, dominant powers dismantle the arts. Native Americans were forbidden to dance; African slaves were denied drumming (but turned it into tap-dancing); armies destroy museums, churches, and theaters to show dominance over ideologies. Yet code language for creative expression emerges somewhere, somehow. Arts and artists *are* social change agents who often lead and even precede larger cultural shifts. They also model innovation, offering effective solutions and experiences outside of the cultural norm. Supporting artistic relevance and creativity in intercultural contexts helps sustain healthy communities and amplifies cultural and global cohesion. And as renowned French-born American cellist Yo-Yo Ma reflects, "If you dig deep enough into any cultural art form, you find humanity."[3]

Engaging your own creativity in embodied dialogue, there is intrinsic integrity. You are already whole. You don't have to go do anything special—buy something or look like someone else—to be more creative. Self-knowledge or self-study is a more generative model: you accept the complexity of the human mind and condition to address and refine your views—what you care about and how that is expressed in your work. In this process, every person is an artist; we shape our lives.

EXPLORING YOUR PROCESS

To get started, take stock. You can only begin where you are. Notice what is actually happening in your body right now—not what you want to have happening but the sensations detailing inner and outer awareness of this moment in time. Creativity requires honesty, truthfulness. This includes a commitment toward not harming self or others in the process of art-making. You make ethical choices when you are working with other people. Clarifying this baseline allows the process of collaboration and co-creation to open. Taking stock of the present moment means you arrive fully, ready to begin—open to inspiration and the process of making.

Find your personal place or pattern for sourcing, locating, and cueing your creative voice. This can be a specific studio, cup of coffee, or eyes-closed moment to enter the process. Questions about this skill arise: How do you let creative impulses flow into expression without too much restriction or convention? What's the pathway to your unique creative flow? How do you get there, know when you've arrived, and find your way back to daily life? Can you recreate the conditions to return to that generative place at a different time, from a new direction, under challenging circumstances or critical views? What do you typically encounter along the way: blocks, moods, or diversions? Pleasures, excitement, awe? As you become familiar with your creative process, those waypoints are signposts, not obstructions.

Dedicate hours in your schedule. The creative process takes time and tenacity. It seems everything conspires to dilute your attention. You generally have to give something up to procure time and to receive the muse of your creative life: cancel a party, ignore email, or skip a trip with a friend. There's sadness in relinquishment, but it makes space for something new. Then you have to get yourself to the studio or writing desk. Clearing your schedule opens pathways to inspiration, but it doesn't guarantee good work: all artists know that you can invite the muse, but you can't make them show up! At first, making consistent time for creative work feels like a hard choice, but eventually the commitment will refresh.

Sustain inquiry and make choices. Although sometimes a song or artwork springs spontaneously forth, the truth is that the process is more like scientific investigation. Many hours of research and experiments with discarded attempts fill the trash basket before the one gestural line in a drawing, cohesive song, or structured dance or improvisation occurs. Focused investment involves both flow and direction. You have a specific goal, but the process of making requires adaptability. Flowing and forming are partners; you need them both to move creative impulse and emotional ideas into grounded form. Visual artist Philip Buller writes: "The needs of the eye and heart have to be in balance. If they are, then the ideas embedded within the painting can emerge as well."[6]

Balance thinking and doing. A life in art is not for the undisciplined. Creative dimensions will lead to surprising aspects of self; fitness and endurance are required. The cortical mind, your top-down brain, likes to have something to do. To keep it from dominating the creative scene, you

Beyond Walls

Making a dance in Iceland, I journey to an artist residency on the northern coast of Skagaströnd for three days.[5] When I enter the main studio, the ten resident artists are all sitting behind computer screens. It has been so windy that no one has been able to work outside for weeks. A works-in-progress showcase is scheduled during the week, so I dive into creating a dance in the building's ice room, made for freezing fish. Blood stains still mark the ceiling and white plastic insulated padding lines the walls. Dancing within these sturdy, storied boundaries motivates my art-making, giving me something to push against in that foreign, expansive landscape. After the evening presentations, the air is still as we return to our housing units. But suddenly the sky pulsates with northern lights—green swirls and streaks lighting up the nearby mountain Spákonufell. Fellow artists, invisible now in the dark, shout across the fields. Spirals swirl overhead as I lay down in the long grasses and watch the sky, humbled. Humans and nature are one. The visual display is beyond art, beyond walls, awesome!

Together

Teaching an embodied writing workshop at Ria d'Etel in Brittany, France, I am reminded that art means "to put together." Assembled are seven women writers from six different European countries, housed at a stone residential center that once held the very last stronghold of Allied soldiers in World War II. German soldiers were ensconced across the tidal river—they could hear each other's voices. The armistice had already been signed, but what to do? Bullet holes in the arched portal to our studio mark this quandary and confusion.

Participants are from countries that were once enemies. For a week, our voices and languages meet, with writing our common thread. Each person excavates stories embodied as memories, feelings that only they can know—some healing, some disturbing. We come together to make something of our shared time in this unique place, to discover what emerges from our overlapping lives and lineages.

Stepping Out

The Celtic Colors International Music Festival in Cape Breton, Nova Scotia, gets you clapping and stomping even if you want to hold still. Celtic music is electrifying, animating, growing right out of the soil and hills as a labor and love of the place. We dine at the fire station on crab cakes and cornbread with locals before heading to the opening celebration. Men and women, children and elders all come together, bringing their talents and cheering each other on. New tunes or fresh interpretations of old ones, fiddling and clog dancing, move the art form forward and inspire the young ones for the future. It is impossible to resist the contagion. It is too much fun!

can give it tasks. First, it can be helpful in setting up a schedule, getting you to the studio on time, and determining a regimen. For example, a session might routinely start with a practice of "arriving," finding ground, space, and breath so all the body is awake and attuned. But creativity is rarely linear. Projects are like a jigsaw puzzle; multiple pieces have to assemble into a whole. Embodied intuitive intelligence needs to take over. All the skills already explored in these chapters are applicable. Throughout, there is a cycling and circulation of awareness. Sensation permeates everything; that's where creativity and intuition begin to manifest.

Absorb, rest, and reflect. Relaxation and integration are essential for specific cellular level physical processes to occur throughout your body. Creativity is not just about producing, "making more stuff." There's a natural rhythm to the process, like breathing in and breathing out, sunrise and sunset, waking and sleeping. Studies show that it's during this time of integration and "not doing" that creativity surfaces. Although studio hours are essential, it's likely that insight comes most clearly on a walk, after a dream, or in some unsuspecting moment when your conscious mind is less dominant, and integration has occurred.

Once your creative process is familiar, any place, any time is for creative making. Inspiration can happen spontaneously. In an instant, you can access a creative state and begin. Sometimes you can spend most of your energy getting started, and other times you are immediately present in creative, focused attention. This embodied state is your personal place of deep, creative connection—available for a lifetime, portable, practical, and free.

PERFORMANCE

There are various ways that the arts touch us, enter our awareness, communicate values and broaden our perspectives. *Transference* is a form of communication that lets someone live an experience through you. When we identify with a character in a film, we "get inside" that person and shift perception. *Transformation* is another: moving or being taken from one energetic and physicalized state to another. We watch characters transform in the process of a play, dance, film, or book. *Ritualization* includes assigning meaning to certain sequences, objects, and practices that can be repeated, passed on from one person to the next, one community or generation to another. We all have small rituals in our days: meals, gestures, or where we sit in a room. Transference, transformation, and ritualization all engage the theatre of our imagination: the capacity to visualize, inhabit, and communicate experiences through reciprocity with others.

Audiences participate in the process of seeing and being seen—the exchange and translation from performer to viewer. Many attitudes exist about this relationship, but most psychologists agree that "being seen" changes what occurs. The observer affects the observed. Performers practice intentionality in this relationship. How close or how far do you want your audience to be? How many viewers, and for how long? Where you place your audience and how you cue your intended interactions have consequences. Architects help shape spaces for effective interaction, offering options for both intimate and spacious connections.

COLLABORATION AND PARTNERSHIPS

Collaboration involves trust. Establishing solid ground with your creative team lets you stay above personal differences and get on with the task. There needs to be some "wiggle room" within personality types and disciplinary skills for intuition and growth so you're not locked into a preconceived way of being within your group. Effective co-laboring is dynamic, reciprocal, and (essentially) forgiving. Although artists may sometimes be considered egotistical and self-centered rather than community oriented, creativity is about relationship, and collaboration is inherent. Behind every lone "genius," there's likely a team of support. You rarely make and present a dance, play, book, film, exhibit, concert, or media offering by yourself or for yourself alone. The work comes to life in the space between the maker and the receiver. And once it's created, it has its own life.

Projects are great ways to get people together who have diverse interests and belief systems. Applying your group effort to clean up a beach, build a children's playground, or initiate a community garden can do a lot to alleviate differences and focus on commonalities. Artists working in group projects know this well; you may not be friends, but when you come together to work, you get something done.

Seaweed Dances
While I'm performing my hour-long seaweed dance with text, a young child in the audience squirms—seeming restless. Later her mother tells me: "she was transfixed; outside she picked up a tree branch from the ground and began moving and swirling, saying 'seaweed.'" And when my great niece comes for the celebratory performance, my real joy is at the end as she brings me flowers on stage and runs out into the empty space and starts dancing. That's the *real* goal of art: to inspire others to make.

France Nguyen-Vincent and Jacques Heim, Diavolo Dance Theater / Architecture in Motion rehearsing *This Is Me: Letters from the Front Lines* (2020)
Commissioned by the Soraya Center for the Performing Arts, Los Angeles, California
Photograph by George Simian

Successful Teams

Over lunch, visual artist Michael Singer, who works on many collaborative projects linking environmental issues with sculptural/architectural designs, says a successful team needs three people: a project manager (who organizes everyone), an engineer (who knows the steps to get things done right), and an artist (who imagines what's never been done before). When working on your own art, however, you have to be all of these on a project. Especially in the early years of your career, you develop multiple skills and learn to appreciate collaborators.

Collaboration

On a Fulbright hosted by Whitereia Performing Arts—a college for Māori, Cook Island, and Samoan dancers in New Zealand—I was assigned to teach choreography to a group of seven advanced students. One woman missed several classes and couldn't finish her first assignment. When it was time to move on to the next project, the students wouldn't continue until everyone had successfully completed the task. I felt that connectivity in their dynamic final performances. Group cohesion was more important than forward motion.

COMMUNICATING ABOUT CREATIVE WORK

One of the challenges of communicating through largely nonverbal forms is speaking about what you've made. Most artists learn to communicate about their work; it is built into feedback sessions and critiques in school, and it extends to grant writing, press releases, interdisciplinary collaborations, reports and assessments, soliciting funders, and sustaining boards of directors. Documentation of your work often outlasts the work itself, especially in ephemeral art forms. Verbal and written tracings require both clarity and curiosity about the process of making; in a way you are describing the indescribable. If you can say it in words, why make art?

Impacts from art experiences may not seem obvious in the moment. Rather, they appear through time. As an observer, it's useful to note that an artwork is the result of months or years of concentrated focus. Your viewing and involvement may be ten minutes for a painting in an art museum or across a span of a few hours in performance. Pause, absorb, contain, notice what stays with you through time. It is the unconscious dialogue that has the most impact. It doesn't really matter whether you "like" or "dislike" something in the moment of experience. What do you remember about a painting, dance, or musical event the next day after dreaming or the next week after living your life?

WHAT WE VALUE

It is curious how we assign value. What's most essential to survival is often overlooked. Like clean air, clean water, healthy soil, and healthy food, the arts provide nourishment and cultural cohesion that's easy to assume will always be there as part of the fabric of life. They are so obvious that they are considered natural and inherent rather than in need of care, cultivation, and protection. Especially in rural areas of developing countries where life is still close to the earth, there is less division between art and life. Yet all of these essential life-components are easily and often destroyed by economic priorities, industrial growth, and political power struggles. The arts are cut from schools, federal funding slashed or eliminated, local financial support diverted to "more pressing" projects.

Spaces for creative making are often on the chopping block. Artists are welcomed in slums, old industrial sites, empty storefronts. Once they create a culture of creativity, economic values rise. Restaurants and stylish cafés come in because there's theatre and dance and diversity. As values rise and rents increase, the artists are kicked out or can't compete and have to move to a different "low rent" district. It's a common occurrence. Soon there are galleries selling artwork but no art-making, no creative energy, and eventually the economy dwindles. Commodification of art is essentially destructive to our communities. How do we value what's essential?[7]

Financial support for the arts, as for all projects, has implications. Who is funding the work and why? In grant writing, you often describe and promise what you will make a year before you do the project. If you receive the grant, you may lose the creative spark just by over-describing. There is a skill to staying freshly open to the muse. It's best to not promise what you don't want to deliver. Commercialism and art-making have rarely been good companions; they have different goals and outcomes. But finding project funding and work situations that establish income streams is part of the process of sustaining a creative life.

As you think of yourself as an artist, you are inside rather than outside the creative process—shaped by the process of making. You both hold an end goal in mind and let yourself change, respond, and be shaken to your core along the way. Surprisingly, the initial impulse for making and the end product are often quite similar, but the process of how you got there will take circuitous routes, and others may not understand (or value) your perspectives. It's like building a house: you have a plan, but then at some point, when the form begins to emerge, the building tells you what has to be done. The work has its own integrity and demands that you listen. You are both guiding and guided. There is risk involved. And there is growth in the journey.

Connections

Today, say hello to your creative and artful self. Rather than hesitating, open to this inherent way of being in the world. Creativity involves paying attention to both inner and outer occurrences. When and how does creativity manifest while you live your life and communicate with others? How do you find time and space today for expressivity and imagination? What lets you expand your range within your chosen medium rather than repeating past patterns? You create your life at every moment, amidst obstacles and distractions. Today, an artistic perspective is your conscious focus, the specific lens of attunement to shaping and making. 👁

TO DO

INHIBITING YOUR INHIBITIONS
60 minutes

Erase the expression "I am not an artist."

Pick an object that you can hold—like a stone, piece of clothing, or your glasses.

- Work in a timeframe—designate one hour from beginning to end.
- Choose a medium to work in: movement, music, writing, visual art, media, or your own combination of doodling, sculpting, and arranging.
- Inhibit inhibition; say "yes" to your creative imagination.
- Holding your object, soften your gaze and open your sensory link to memory and associations.
- Follow points of inspiration or connection without judging or second-guessing yourself; it's an exploration.
- Make something, giving form to your imagination.

Being Inspirable
Choreographer Paul Taylor, in his film on the creative process, describes walking by a trash bin on his way to rehearse a new piece with his company and finding a discarded CD that would become the music score for the dance. Being inspirable is a conscious state. We can cultivate it in ourselves and in daily life. Rather than shut down, we open up.[8]

Hope
Playwright Tony Kushner writes in the *New York Times*: "It is an ethical obligation to look for hope; it is an ethical obligation not to despair."[9]

Random Works (2019) by Gordon Thorne
A.P.E. Gallery, Northampton, Massachusetts
Photograph by Stephen Petegorsky

TO WRITE
INSPIRATION
10 minutes

Imagination and inspiration are linked. Respond to one of the art images in this book or to a familiar artwork in your home. Let it be a jumping-off place for your writing. Use embodied writing, feeling what you are saying. At some point, stop and move; then write more. Notice if moving changes your writing.

IV
ENHANCING
AUTHENTICITY

Shayla-Vie Jenkins in *Discourse*
Photograph © Anna M. Maynard

Face and Expression

Let's face it: the face is a signature, an indicator, and a communicator. We cue ourselves and each other with our faces through language and expressions. Embryologically, it develops with the hands and feet—not with our central core; it is part of our periphery, like fingerprints. The energy of the face is impacted by our thoughts, feelings, and state of attention; there is a mental-emotional connection that shifts whole-body physiology. Change any one part and body chemistry is affected. Lift an eyebrow, turn down the corners of the mouth, flare the nostrils, and it shakes everything up—hormones redirect. We play our moods with our faces, like fingers on a piano.

We look at our face often; it tells our life story. We present a forward-facing visage to the world, but what is actually going on inside ourselves—our inner experience? Where we hold stress often shows up in the face, particularly around the eyes, jaw, and forehead. Many words, images, and descriptors impact self-image: we can feel sad-faced, shamefaced, or effaced. Saving face offers relief, whereas being two-faced gets us in trouble. Sometimes it helps to just take things at face value. And in complex situations, there is the simple request: Can we just talk face to face?

What is the energy we want to project through the face in various interactions? To enhance skillful communication, we can revitalize the face muscles by stretching, strengthening, and relaxing. Exercising the face helps to reduce undue tension just like full-body workouts refresh and energize. Oxygenating the underlying muscles stimulates connective tissues that help the skin to be firmer—linking all the way down to the feet. As muscles become both stronger and more responsive, a neutral resting state is restored. Relaxation of tension brings fresh energy, brightens the face, and creates congruency throughout the whole body.

FACTS

Each person's face is unique, yet the embryological origin of all human faces is the same. Facial development begins at the end of week four during embryogenesis; you have a head at week seven, and by week nine, the roof of the mouth (hard palate) is complete.[1] Special senses move into their locations by birth, with the sides of the nose, lips, and upper palate meeting along the midline if all goes well. Facial evolution continues after birth: at puberty the sinuses develop and define the adult facial imprint. The mature protective skull is an interwoven puzzle of twenty-two bones, offering both stability and mobility to protect the underlying brain. Fourteen of these pieces create the basic "look" of the face, with the maxilla a centerpiece structure linking the eye sockets, bridge of the nose, and upper jaw.[2]

Nonverbal communication through the face speaks volumes. Over forty-three muscles orchestrate our expressions, the most of any species.[3]

What clouds your face?— Chimamanda Ngozi Adichie, from Purple Hibiscus

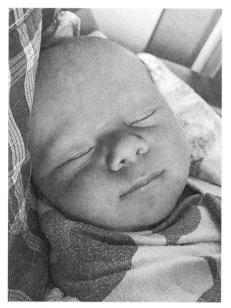

Ewan Elijah, newborn
Photograph by Jonah Keith (2020)

Smiling

I was raised in the era of Little Miss Sun-
beam, the Jantzen Smile Girl Contest, and
Miss America pageants. As I participated
in each of these as I grew up, the smile was
everything, along with the accompanying
body posture—chest out, hips back, torso
angled toward the judges. Similarly, in
ballet classes épaulement involved turning
your head, neck, and shoulders from the
waist so that you were facing the audience
in every position—originally to show
deference to royal viewers in court ballets.
When I traveled from the Midwest to New
England to study modern dance, I had to
learn to "unsmile." We were *serious*! The
whole body was expressive, with different parts emphasized in each technique.
Foremost was the pelvis (Martha Graham),
then the torso (José Limón), every isolated
joint along with lightning-fast feet (Merce
Cunningham), the thighs (Paul Taylor),
and the fingers and eyes (Madame Shan-
kar). After all of those immersive trainings,
finding a relaxed face was not so easy.

That's a lot of muscle activity in a small space, allowing over ten thousand
facial configurations! We are programmed to relate so we can work together, get along, survive, and thrive. Rings of muscles surrounding the
eyes and mouth create sphincters that open and close. Other muscles pull
skin and facial features in every direction—up, down, or out to the sides
of the face—in different degrees and combinations, creating multiple expressions. The parts can work together or independently: one eyebrow
can be raised while the other remains fixed. Intricately connected to our
emotional limbic brain, the face muscles reflect our mood and emotional
state unconsciously, serving as an inherent method of communication that
begins during interactions with our mothers.

Maternal-infant connection and bonding depends significantly on
facial expressions. We read so much from babies' faces while they are
reading ours. All this occurs before sounds become words, linking hearing and speaking as a next layer of communication. Researchers say that
an estimated twenty-five thousand hours of interaction with caregivers in
the months after birth create deeply embedded habits that involve mirror
neurons in the brain, which facilitate behavioral imitation. Inhabiting the
face freshly sometimes requires relaxing these programmed responses:
moving beyond pleasing, affirming, defying, or complying to a more relaxed, nuanced, and responsive palette.

It is generally agreed that there are at least six common emotions
shared among all humans. Understanding the micromovements of *affects*,
which are inherited biological responses, in relation to *emotions*, which are
learned responses, helps clarify and refresh behavioral choices (see Day
20). Affects are programmed as reflexes before birth and are facial patterns
shared by all peoples, whether living isolated in a rainforest or immersed
in a crowded urban center.[4] These foundational but fleeting responses include surprise, sadness, happiness, fear, anger, and disgust—the inherited
biology of emotion. Affects last only a few seconds, a flicker in time, before
conditioned emotional patterning takes over, reflecting social and cultural
norms. Personal facial expressions, embedded during the context of a life,
may function as deeply ingrained and seemingly involuntary habits but
can be refreshed with attention.

Home to the *special senses*, as well as thousands of nerve endings, the
face creates a sensitive map for active exchange. The high nerve density
means that there are not only many possibilities for movement, but also
the face-brain connection is especially sensitive to both what we're expressing (awareness of our own face) and how closely we pick up other's
expressions (awareness of others' faces). The energy of other people enters
our personal energy field through the receptors of the face. The tongue
and lips join the fingertips as exquisitely sensitive parts of the body. The
trigeminal nerve (fifth cranial nerve) is the largest and most complex of the
twelve cranial nerves. It is responsible for sensation in the face and motor
functions such as biting and chewing. The *facial nerve* (seventh cranial
nerve) also enables facial movement; the main root emerges in front of
each ear with branches activating the face, scalp, and outer ears. Neurons

interact with multiple brain centers, both sending and receiving messages in relation to inner and outer stimulation. Lower face muscles are more fully represented in the sensory/motor cortex of the brain than the upper face, enhancing refined articulation of the mouth for speech as well as the sensitive, expressive dance of our lips.

Facial features have particular symbolic and metaphoric connotations. Along with being generously endowed with movement possibilities, the parts impact a person's body image: the hue of their eyes; shape of their lips; contours of their nose and cheekbones; color, texture, and markings of their skin; and the underlying web of fascia that weaves the parts together (including dimples and wrinkles), creating a whole. To refresh the relationship to this face we call our own, we can clear unnecessary memories and ideas that limit expression and resilience through conscious attention. Reinhabiting the face helps us to appreciate the parts and all they do for us at conscious and unconscious levels.

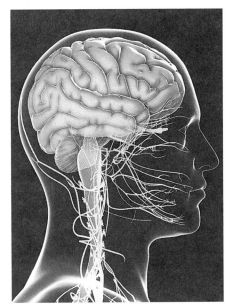

Nerves of the face
Graphic © Sciepro / Science Photo Library

APPRECIATING THE PARTS

The *nose* is the oldest of the special senses. Evolving through natural selection in fish, the nose literally leads us through space. As part of a bilateral organism with head and tail "ends" connected by a spine, the nose detected chemical "scents" in the water for survival choices: move toward or away. In humans, scent registers directly in the emotional-relational limbic brain and is linked indelibly to memory. The scent of your grandmother's kitchen and your child or friend are part of affiliation and bonding. The smell of anger, danger, and changing conditions can save your life. Mostly unconscious, background scent can become foreground in an instant: when the dinner in the oven starts burning, take action!

The contours of the nose describe the landscapes where our long-ago ancestors lived. The size of nostrils and the width of the bridge are encoded in our genome. A thin Nordic nose preserves heat and moisture in cold dry climates; a broader, more spacious nose that cools down the body amidst humidity and heat is essential near the equator. As well as locating us in time and place with its shape and connection to memory, our nose filters the air of harmful materials with its inner hairs. The nose is often a visible indicator of health and emotions. A dripping or running nose can signal a cold, allergies, or emotional upheaval. Because the tear ducts (*lacrimal glands*) of the eyes drain into the nostrils, we blow our nose after a good cry. Metaphorically, for English-speakers being "nosy" implies we are invading other people's affairs, and we might be admonished: "Keep your nose out of my business." Or if someone is upset or irritated, we might hear: "He got his nose out of joint." As a symbol, the nose has multiple meanings, depending on cultural and ideological interpretations.

Lips offer a choice: to take in or not. Although it may seem when you are in love that they are mostly about kissing or smiling, lips also lead to the dark cavern of the mouth—teeth, tongue, and saliva for eating and talking. As one of the most sensitive parts of the body, the lips determine if we speak or remain silent and to what degree. Some days your "lips are sealed" even

Greetings

For the Māori traditional greeting in New Zealand, you press nose-to-nose. Standing in a circle at the spacious University of Auckland dance studio, we go one by one around the group. The custom, called a *hongi*, involves sharing breath. Forehead-to-forehead, nose-to-nose, it is intimate; you are no longer a mere visitor—you are a *tangata whenua*, accepting responsibility for your actions within the land and toward its people. I can still feel the heat and pressure of encounters, remember the breath of exchange.

if there is much to say. Other times words and sounds tumble right out of your mouth as songs, screams, or sighs. In certain situations, you "bite off more than you can chew," "remain tight-lipped," or "mouth off" at inappropriate moments. Keeping your mouth shut and resisting showing your teeth are aspects of expression monitored at the lips. There's a lot of tension and attention at the sensitive, tell-tale entrance to the mouth.

The *tongue* articulates. It is a muscular organ in the mouth that is essential for speech as well as maneuvering food for mastication and swallowing. It is the primary organ of taste within the digestive system. In the back of the mouth, the tongue is anchored into the hyoid bone and links to the skull and thorax through connective tissues. Covered with moist, pink tissue called mucosa, the tiny bumps called papillae give the tongue its rough texture. Thousands of taste buds cover the surfaces of the papillae that link to the brain, which recognizes four common tastes: sweet, sour, bitter, and salty. The tongue is a sensitive, sensual part of human anatomy utilized at all ages; babies explore by putting things in the mouth; kids stick out their tongues in concentration or to show dislike; and adults kiss, drink, and dine—savoring the flavors of each. Metaphorically, the tongue is associated with language and expressivity.[5] Your "mother tongue" is the language you grow up speaking—your home language; in some multilingual families this can include a variety of sound production requirements and their associated meanings. "Holding your tongue" is a common expression in English for remaining silent. Relaxing or stretching the tongue can assist with release of stress; in some yoga and meditation techniques, the tip of the tongue is rolled upward in contact with the roof of the mouth to affect energy circulation and mental states.

The *eyes* direct us here and there. We position the body to see and be seen. Looking inward, looking outward, the eyes are not at all objective about their goals. As with other pairs of special senses that arrange themselves in the womb, there's both a joining and a staying apart. In fetal development, the eyes know just when and where to stop. The *eyebrows*, as arched observers overseeing the whole, indicate the exact tone of expression. Central in cartoons, the eyebrows slant up to reflect excitement, down for a furrowed brow, or remain horizontal for a quizzical, undecided, or skeptical demeanor. Try moving your eyebrows, and you'll feel the significant impact of their positions.

The *face* can be repositioned in any direction by the muscles of the head, neck, and spine. The multidimensional agility of *Homo sapiens* is one of our survival characteristics. Both in the structure of our musculoskeletal system and the complexity of our nervous system, we are made for movement in all planes. With a distinctly unstable, high center of gravity balanced over the small base of our feet, we require neurological responsiveness. We alone as a species (as far as is known) can think obsessively about the past, present, and future. Unless we rigidify in mind and body, we are endowed with multidirectional possibilities as embodied beings. Physically, we can spiral around our spine to change facings, spin like a martial artist, and flip, turn, twist, bend, and contort to position our special senses where we most want to direct them.

Masks and make-up can amplify or disguise facial features according to intention. External masks, normally worn on the face, are utilized ceremonially for invocation or protection as well as contributing to disguise, health, performance, or entertainment. When you change the portal of your face, your whole body responds, including emotions. Halloween and the carnival festivities of Mardi Gras let us experience the ways our fingers and spine transform into a witch, cat, or goblin as called forth by our mask. We can create internal magical masks as well, taking on the characteristics of another person's energy, body shape, and time period as actors or simply to deepen understanding by "walking in another person's shoes."

Cultural implications of the face and its various coverings and adornments signify intention. Many people globally wear some sort of make-up or cosmetic preparation to "put on a good face" before going out in public or facing challenge. Appearing without some of the many products on the international market amplifying, accentuating, highlighting, or hiding the facial features can lead to anxiety and vulnerability. Facial hair and skin treatments accommodate custom or personal goals, reflecting individual perspectives around belonging and expression. A sense of self-identity often comes through the prominence of the face in communication. Traditional and contemporary creative masks, make-up, facial adornments, and markings can help us see deeper into layers of meaning in life. They indicate how we fit in culturally and how we choose to present ourselves in our community.

Jero Luh (2020), mask
Ida Bagus Anom Suryawan, artist
from the village of Mas, Bali, Indonesia

Disguise
Wearing masks during the coronavirus outbreak, we stop at a local farm to pick up food. The mandate is to stay six feet apart for social distancing; the joke in rural regions of the state of Maine is: "For Mainers, that's pretty close."

EXPRESSIONLESS

Signals to the face can be interrupted in challenging circumstances—expression is put on hold for survival. Being expressionless is different from relaxation; it is protection. No one can know what you are feeling or thinking until it is safe enough—maybe never. War, abuse, shame, or threat can activate the sympathetic nervous system to suppress communicative responses in order to protect you from aggressive behavior from others. When you are helpless to defend yourself, a noncommittal mask can save your life. Rehabilitating responses can take some work. Ingrained habits are hard to change.

REHABILITATING THE FACE

Three foundational cues can help facial ease and authenticity: (1) *Relax the jaw*: When the jaw is relaxed, there is less stress in the body. If there is less stress in the body, the jaw relaxes—an ongoing feedback loop. (2) *Feel your feet on the ground*: When you have foundational support, the face can be free from holding you "up." And (3) *Move your face muscles each day*: When you explore inherent expressivity, you release blocked tension and energize the whole body. You can use your face both to elevate your own mood and to increase connectivity with those around you.

What's Missing?

When I had surgery on my nose (skin cancer from too much sun), a big white bandage drew focus for several weeks, like a beak or beacon. My eyebrows and lips wouldn't move; even brushing my hair hurt. My face took on new dimensions. (My friend Annie called it my snowman face—a carrot for the nose.) When I couldn't respond to passersby, they looked away. What we leave out of a project reveals its limitations. All this focus on my face made me recognize that I'd written twenty-one chapters for a book on communication over the last five years and included nothing about the face. Really! We are drawn to what we need to learn, whether we want to go there or not.

We meet the world with our face. When we change our face, we are altered from the inside out. The face is a nonverbal communicator (to ourselves and others) as well as the home for speech. Recognizing that the skin of the face connects all the way down to the soles of our feet and the webbing between our fingers reminds us that the body is one embodied organism. This interconnecting skin is both a barrier and a bridge, and the face is one highly adaptive and sensitive portal for communication. Feeling that all the parts are interwoven, we can face life as it is with our whole expressive bodies present.

Connections

This strand of the tapestry of embodiment focuses on the face. Noticing your face throughout the next twenty-four hours offers a lens on expressivity and authenticity. When and where do you smile, frown, cry, or find a neutral, easeful face? Does the tension level in your facial muscles match your inner world or mask it? Is there a place where you can relax and feel most authentic? The face is an indicator, a communicator. What you feel and what others see in communication may or may not be the same. Sometimes it is helpful to give your face some attention beyond the usual conversation with the morning mirror, noticing what that expressive landscape is up to and refining!

TO DO

EXPLORING YOUR FACE
10 minutes

Let's face it: the face is personal.

Preparation: Wash your hands, sit comfortably, and take a few deep breaths.

- Let's begin with the forehead: Using both hands, take a few finger strokes side-to-side to clear tension from this important area. Wipe away tension.
- Bring your thumbs to the undersides of the eyebrows, below the boney ridge. With gentle but firm pressure, stroke outward releasing muscle tension.
- Stroke down along the outer nose (sinuses), under the cheekbone ridge, and out toward the jaw joint in front of the ear hole. Repeat a few times.
- Touch the indentation under your nose and stroke outward to clear the upper lip region.
- Place your thumbs under the chin, and stroke outward and upward toward each ear, clearing the muscle attachments that may feel lumpy and bumpy from tension.
- Massage around your ears, and gently palpate the fleshy parts of the upper ear and lobe.

- Stroke down your neck to drain tension.
- Then locate the "trough" or indentation above each collarbone; starting at the center, stroke outward toward the shoulders and massage any blocks or stuck places. This is an area where lymph drains into the heart like a gutter: you don't want it to be clogged.
- To finish, tap around the face and skull to enliven circulation. Massage the scalp from the forehead to the back base of the neck.
- Sit quietly and add a slight smile; notice its effects on your whole body. Do the same with a slight frown and feel its effects.
- Find a neutral position and notice the sensations of an enlivened face and skull. Repeat daily to keep the facial tissues enlivened and healthy.

TO MAKE AND WRITE

MAGIC MASKS
60 minutes

Masks are powerful tools to evoke, amplify, and personify. For many, protective face masks have become part of daily life in response to airborne health challenges and require new communication strategies. Tap on the doorway of your imagination: If you could wear any mask and transform into its character, what would you choose? Imagine or create a mask from materials around you. Be creative—leaves, grass, paper, oatmeal, paint, and feathers. Put on your mask and write from that persona/character. Let it affect your whole body, your voice, your tone—experiencing "other." Let your character emerge as a larger part of yourself. Take a selfie and write about your experience. If you want to continue exploring and reflecting: What was your experience of masks in your past?

Reminder
Mevlâna Jalâluddîn Rumi, the thirteenth-century Persian poet, reminds us: "Whatever lifts the corners of your mouth, trust that."[6]

Day 17

Energy and Vitality
With Rich Wolfson, physicist

We exist in a bath of vibrations and so do all living creatures. — W. H. Hudson, from The Speaker

Wild Ponies

As we hike on the Devon moors, the wild Dartmoor ponies keep a wary eye. A sign describes these hardy companions: "Tough, sure footed with fabulous temperaments, they can survive the harshest of moorland conditions." Their hoof prints are all around as we stroll over the rolling landscape toward peaks called tors breaking through the heather. At one point, we stop to dance, meeting the energies inside and around us. The ponies come near, interested. This way of communicating seems familiar and non-threatening: energetic sensitivity is part of wild nature—including ours.

Energy is that invisible substance that activates human life from first breath to last. We cannot see it or touch it, but we feel its presence moving in our bodies. When we have too much energy, we knock things over; too little and we cannot get ourselves out of a comfy chair. Energy also surrounds us in the air, water, plants, animals, soil, ancient rocks, and mineral crystals—in every animate and inanimate presence. We watch plants grow from seeds, animals frolic in the fields, and feel gravity pull on our bodies. Energy is measurable (as motion, heat, or electromagnetic fields), yet it is formless and takes up no space. We recognize energy because it moves through form: we see the wind in trees, feel motion in our bodies, and ride waves on the surface of the ocean. Energy is a mysterious substance, but it is fundamental to our lives and liveliness.

Matter and energy often exist together and change states. Fire, for example, consists of visible matter (in wood and gases) plus invisible kinetic and thermal energy: the burning wood yields heat.[1] Thoughts, dreams, and emotions are invisible, yet they create substantive physical changes in our bodies through neurochemical reactions. Transformation occurs as one state or condition changes to another: a space of "betweenness." In other examples, water as frozen ice melts to fluid; when boiled it becomes steam. Food is broken down in our stomachs and intestines and is transformed by the mitochondria in our cells into fuel for our bodies. Waves move across the ocean's surface, yet the water itself remains largely in place until the wave hits the shore.[2]

Professionals who carefully monitor and regulate their energy levels—like surgeons, performing artists, and competitive athletes—become masters at precise calibration in the moment. Sustained attention is a practiced skill, along with the capacity to refine and adjust energy expenditure in relation to context. Energy plays a powerful role in communication and interaction: how much you commit, how present you are in the process, and the ways in which energies are exchanged. Both inner awareness of self and outward connection to others are necessary.

FACTS AND NAMING

There are many names for the invisible energies inside and around us, and we cultivate connection to those energies and make relationships with them through language. Linguistic identifiers are one place where physical sciences, social sciences, and humanities (including religion and the arts) have differed and even conflicted in the past. Yet with current physics assuring us that vibrating particles establish the basis of all forms, we can recognize that what is perceived to be solid is actually in motion—even the

Voyage (2018), Diavolo Dance Theater /
Architecture in Motion
Artistic director Jacques Heim
Photograph by George Simian

chairs and tables where we sit. In this notion of invisible energies, science and mysticism meet. Embodied creative practices affirm that humans are part of a universal energy field that surrounds, binds, and includes us.

From a scientific perspective, we learn that energy exists in several forms. These are identified in the English language as mechanical, electromagnetic, and nuclear energy. *Heat*, or thermal energy, from the movement of atoms and molecules, relates to temperature. *Kinetic energy* is the energy of motion, like in a swinging pendulum. *Potential energy* involves position. For example, a ball sitting on a table and a person balancing on one foot have potential energy in relation to falling. *Mechanical energy* is the sum of the kinetic and potential energy. *Electrical energy* is associated with electrically charged particles (electrons, protons, or ions). Movement of electrically charged particles results in magnetism and *magnetic energy*. Together, electric and magnetic energy join in a dance to make *electromagnetic waves*, which include light and carry energy throughout the universe.

From cultural and religious perspectives, energy has broader meanings with differing linguistic signatures. For example: in China life force energy is called *chi* or *qi*, in Japan *ki*, in India *prana*. *Mana* is spiritual energy in Polynesian cultures, which can be inherited or bestowed. In Spanish, at least four nouns discern types of energy: *la energía* (power, drive, spirit), *el vigor* (vigor, lustiness, zip), *el* ñeque (strength, vitality), and *la fuerza*

Susanna Recchia, London, England (2015)
Body and Earth: Seven Web-Based Somatic Excursions
Still image from videography by Scotty Hardwig

Waves

Standing at the Asilomar Park beach on California's Monterey Peninsula, physics colleague Rich Wolfson says to me, "You know, the wave is not the water." We have had exchanges before where Professor Wolfson's perspective shifts my way of seeing. Co-teaching the course "The Dance of Physics" several years before, we cultivated congruent perspectives. "What do you mean?" I respond, and he explains. Although I remain baffled, I make a dance with this view in mind and include text: "A wave is not the water. It is just energy moving across the surface; the water molecules remain largely in place until the wave hits the shore. The mind is not the brain. It is just energy moving through the tissues; the cells of the body remain largely in place." Embodying the words helps me understand: invisible energies manifest in form. It is a perspective worth understanding.

(force). In Norway *friluftsliv* is the energy of oneness between humans and nature that can restore balance among living things. Within the richness of diverse perspectives and differing religions, energy is often linked to God, gods and goddesses, and powerful mythological beings or unknowable presences. In Taoism, for example, this primary energy field is *the way*, the original source of life.

The first steps in self-managing energy levels include sensing and feeling the existence of energy moving through the body and using language to accurately describe the sensations. Energy in tissues can feel like heat, vibration, pulsation, or tingling. It often is a blend of activated sensations and sometimes results in trembling, twitching, or shaking. Motion is involved. We know when our energy is low: we feel drained, lethargic, uninspired. At other times, we feel energized, ready to engage. Some people learn to wear off excess energy (like going for a run or bike ride) before interacting with others so as to not overwhelm them in conversation. Others have to rest and build energy reserves before socializing. Homeostasis, or balance, is the body's desire. Returning to a balanced state easily and quickly is a sign of health after a necessary burst of energy in a lightning-fast encounter or confrontational interaction. The words energy and power are often used interchangeably (although a physicist would insist on respecting their difference) and are associated with force, strength, might, and productivity.[3] Yet fluidity and calm are equally essential. There are useful practices to enhance energy regulation, helping us stay in balance: not over-energized or depleted, not uncontrollably vacillating from low to high energy, with extreme peaks and falls (see Days 26–30).

CHOICES

In the body, we use energy in multiple ways. All parts of the body (such as the muscles, brain, heart, and liver) need energy to work. This energy comes directly from the food we eat. The energy source in food is stored energy from the sun processed by plants. Peanuts, hamburgers, and chocolate, for example, provide their nourishment through the roots of the peanut plant, the leafy stems and grain crops that feed our beef, and the cocoa trees and sugar cane that combine to sweeten our chocolate treat. Breath moving into the lungs is another primary place where our internal energy meets the energy of the environment. This requires plant life to produce oxygen and absorb the carbon dioxide we produce and to photosynthesize food energy. Movement is a third energy-building resource; we move to feel ourselves moving, breathing, and circulating essential nutrients throughout our systems. Movement involves kinetic energy; it creates more movement. Prolonged sitting, on the other hand, can diminish and drain energy levels and vitality—there is a lack of nourishment to the vital organs. Illness may take over, forcing us to lower our energy level while we recover. For some, feeling energized means being busy at every moment: a human doing, rather than a human being. Becoming comfortable with restful energies and calm states can enhance and balance a life. This allows receiving as well as giving, experiencing reciprocity between self and other.

Most people are bound to ideas and feelings about their bodies. This can include a self-image that reflects personal energy levels—what feels good or natural. She is "high energy," he is "chill," or they seem to have "boundless" or "chaotic" energy when they are together! Identity descriptors can be based upon what you have been told, seen reflected in the mirror and in photographs, or what you feel from other people's responses. Without clear principles and practices for working on energy modulation at different stages of development, it is easy to amplify stress. Instead, we can track subtle energies that may be held, blocked, or restricted from flow and have a more flexible and responsive body image in interactions. To modulate our energy levels, we can consider daily: What makes us feel good, most like our genuine self, most helpful to others? What people, places, and activities support rather than diminish our vitality?

As we recognize energy within ourselves, we can also notice patterns in others and make choices about interaction. "Positive energies" reflect mental attitudes including interconnectivity, appreciation, and good will, while "negative energies" often feature insecurity, competition, and aggression. Animals are indicators of energy states in humans. Horses come toward or stay away; your dog hides in a closet when tense conversations escalate; your cat snuggles onto your lap or chest when you are sick or sad. We can engage this quality of intuitive intelligence to recognize when to open to other people's energies and when to self-protect. We can practice both opening toward and defending against—establishing self-protection through our own healthy psychological and physical energy. In qigong, for example, exercises help us build *wei qi*, the protective field around the body.[4]

Dancing Together

When my husband had heart stents inserted, he was required to take beta blocker medication to reduce adrenaline flow to the heart as it healed. Our joke was that this made him so lethargic, he would do anything I asked—including taking swing dance lessons. "Follow the lead, wait for the signal from your partner," our patient teacher would call out to me. I had to hold back, and he had to firmly initiate. We would bump into each other, twirl backwards, step on each other's toes, and hear the beat differently. We gradually found our groove. Now (without medication) we enjoy our private Friday night kitchen dances: nobody but us knows what we're up to.

Emergence-y

We have just planted seeds for our indoor garden. It is the end of March and still snowing in Downeast Maine, but we are hopeful. Days grow longer (in this easternmost part of the U.S.), and sun fills the windows. We are in an initial phase of the coronavirus pandemic, and growing our own greens seems like a good idea. The plants, like us, are trapped inside for now. I look each morning to see if pale green shoots have pushed up through the potted soil, anticipating parsley, basil, and baby kale. Plants: we can't live without them. They fuel our every breath and movement. They also calm our moods.

Storm

Wet snow makes heavy branches. During the night storm, trees break; one just misses our front porch. In the morning sunlight, we survey the damage. A branch of one tall spruce tree rests on the power line, making it sag. Over a hundred thousand people in the state are out of power; we don't want the line to break. The long icicle dangling from the branch hangs over the wire, perpendicular to the earth: it is bound to fall. The term *potential energy* comes to mind; gravity reigns. I watch all day as sunshine slowly melts the overhang: drip, drip, drip. When I turn away from the window, it drops. The branch springs upward and away from the line. Energy is demonstrated on several fronts—including my sigh of relief that we still have power.

Enthralled

On a solo performance tour in Beijing, China, I am driven from the National Ballet Institute to the Cultural Center. This is an unplanned opportunity to work with Mongolian dancers newly arrived from the provinces. In the dark basement room where we gather, the dancers wear boots and heavy clothing for outdoors, their bodies still primed for the vast expanses of their homeland. As drumming begins, energy is literally ricocheting off the walls. We improvise together: first one joins, then another, and then the whole group. Energetic is hardly a sufficient word to describe my experience; enraptured, enthralled, en . . . meshed. "En-" is a prefix meaning "to cause to"; it changes adjectives and nouns into verbs. We were all verb, reverberating.

Energy is everywhere, both inside and outside of our bodies. Since energy itself is invisible and formless, we can recognize that energy is always carried by something more tangible; it moves through form. Sensitizing ourselves to the energies within our bodies lets us self-manage our stress levels and enhance vitality, noticing sensations like tingling, trembling, and waves of heat. Acknowledging the energies of other people, places, and changing conditions helps us make wise choices in embodied communication. The same energy can engage a conversation that brings joy, or it can go toward hurting yourself, others, or the world—fighting with a friend, making negative comments, writing untruths that damage and cultivate distrust. The illusion that energy is beyond our awareness is outdated. We now know that energy and mind are the places where we have influence: we can change how we feed our energy system throughout the day and direct our intention toward enhancing our lives and bringing well-being to others.

Connections

What does your energy feel like today? Start with this baseline, and then chart your energy levels over the next hours and notice what fills or drains your energy reserves. As part of this process, consciously intend to include a consistent centering practice at different intervals, reinforcing your deepest knowing and authentic self. Choose one, such as the meditative practice of noticing breath or the modulating tone exercise in this chapter and include it in your routine. Energy and attention are linked through intention. Like brushing your teeth, engaging in an effective practice just a few minutes each day can impact health and longevity. Then, in stressful situations, you are prepared and "rehearsed" for less reactivity; your energy levels are refined to be attuned to the moment rather than to under- or over-exert.

TO DO

MODULATING TONE
10 minutes

> *We each have the full range of energy potential, from the most delicate touch of finger to a baby's face to wild, free dancing. Explore this range so you can become more comfortable with tonal shifts in your body when communicating.*

> Standing: Begin with any movement that feels good to you—walking, rolling, dancing . . .

- Now, imagine you are turning a rheostat—a lighting dial that lets you control the brightness or dimness.
- Move with your energy dialed to low. How does this feel in your body?
- Turn the dial and continue moving at medium energy.
- Turn the dial further and move at full, dynamic energy.
- Then play, modulating the energy levels and physical tone in your movements.

TO WRITE

HIGH-VOLTAGE WRITING

10 minutes

In detective writing when you are learning how to get a narrative going, there's a saying: when in doubt, have a man come through a door with a gun in his hand.[5] Write an anecdote (a short story with a beginning, middle, and end) about a time when your high energy was called to the foreground. Allow yourself to reinhabit the moment, speaking from inside the experience. Bring something fierce into your writing, a game changer. Read aloud and notice how it feels to embody high-voltage writing.

Energy Bunny

"You are an energy bunny," my neighbor in Florida calls out as I zip by on my bicycle for the third time that day. You have to realize, I'm seventy-one years old and this is an age fifty-five-plus trailer park by the Atlantic Ocean where I spent my winters as a child. Headed to the ocean I still feel like my twelve-year-old self: breeze to face, legs pumping, laughing. My ninety-three-year-old neighbor who knew me back then says, "You never did sit still, you were always in motion." Energy levels are with us from childhood, part of our identity in the world. No bouncy ponytail or smooth skin accompany that description of my vitality now, but I'll accept the acknowledgment of youthful spirit.

Exchange

On tour with college students, I surprise them by often having more energy than they do, more endurance and focus for work, and more left over at the end of the day. Part of this is from years of performing: don't waste your resources—financial or physical. Channel your energy for what's important, and don't let others distract you. There are so many ways we drain our energy all day long. A Māori friend explains that his deep soul energy is called *Hatu*, his spirit. You share your energy, but you never give away or drain your *Hatu*—or there's nothing to give.

Day 18

Humor and Laughter
With April Danyluk, former Masters student in Public Administration

Learning to Be Funny

Tula Isabel, as a young child (five) with an older brother (ten), would try to tell us jokes. She knew they were meant to be funny but didn't understand the process. She would tell the same joke over and over, make one up that had no punch line, or resort to bathroom jokes. Frustration would set in when there was no laughter from captured listeners. Humor, as with language, is a learning process. Sometimes she would get a good one: "Why do cows wear bells? Because their horns don't work." We all laughed and laughed. Success!

Humor saves us—literally. A mirthful laugh releases hormones in the brain that relieve and revive. Like a good shower or a plunge in a pool on a stressful day, well-intended humor combines surprise with fun. Ha! A friend with a sense of humor can cut through tension. Clowns, tricksters, comics, and cartoons broaden our perspectives, letting us laugh at ourselves and our human situations. Dark humor, like all human endeavors, can uplift or destroy. Making someone the butt of a joke, poking fun, or cutting someone down may send a knife to the heart—intentionally diminishing them. We have choices about how we use humor in our lives. Recognizing its potency, we can cultivate proactive aspects like hilarious encounters and outrageous levity, letting ourselves have fun and be funny. For a bit of relief, make faces, jump, skip, twirl, shake out your body, sing at the top of your lungs, or laugh twice a day for no reason at all.

We are born to laugh. It is a natural, nonverbal way of expressing. Laughter involves releasing excitation in the nervous system by an expulsion of breath and, sometimes, sound. Watching a four-month-old baby giggle, smile, and laugh, we recognize that laughter is inborn and universal. Babies cannot yet speak, but the brain utilizes laughter to communicate both with itself and with others. For example, if there is a loud sound— like the parent spontaneously sneezing while the baby is strapped into a highchair—first there is a bit of a jump (surprise escalates to startle, eyes open wide, and super alertness activates). Then—if there is no danger or pain—explosive laughter and smiling follows (enjoyment to joy!). The child finds it absolutely hilarious.[1] (Crying is another option if the situation is perceived as disturbing.) When we realize that a surprise is not dangerous, we laugh because we are relieved. Even when the event is repeated, like being startled by a balloon popping, it is still so funny. There is both a discharge of the initial excitation and a reaction to one's own reaction; it feels surprising when we "jump" for no reason. The degree of safety and security can determine the difference between laughing and crying.

Well-intended laughter is healthy. Gelotology is the study of laughter and its effects on the body. Derived from the Greek γέλως (*gelos*: to laugh, laughing, laughter), gelotology was established as a field of study in the 1960s to research and communicate the effects of laughter.[2] This includes both the physical act of laughing and the playfulness or humor that causes it: physiological and psychological perspectives. One of the pioneers interested in the physiological benefits of laughter was Dr. William F. Fry, a professor emeritus at Stanford University. According to Dr. Fry, "muscles are

Peter Schmitz in *Stand Up* (2015)
Photograph by Peter Raper

Dangerously Funny

In her book *Clock Dance*, author Anne Tyler describes the overwhelming and embarrassing sensations of nervous laughter taking over your body: "These giggles were like a liquid that flooded Willa's whole body, causing tears to stream from her eyes and forcing her to crumple over her carton of things to sell and clamp her legs together so as not to pee. She was mortified . . . but at the same time it was the most wonderful, loose, relaxing feeling."[5]

activated, heart rate is increased, respiration is simplified with increase in oxygen exchange—all similar to the desirable effects of athletic exercise."[3]

Research has brought laughter into the matrix of therapeutic techniques for managing pain and treating conditions such as depression. Continued in-depth study confirms that laughter impacts every system in the body. Significantly, there are seven recognized benefits of laughter: it releases physical and emotional tension, improves immune functioning, stimulates circulation, elevates mood, enhances cognitive functioning, and increases social engagement.[4]

Inner laughter can have similar benefits to outward laughing. For those of us who are less publicly expressive, when we perceive something as funny, only a snicker or a silent chuckle may emerge. Yet we receive similar physical effects as from boisterous laughter, including an after-flush period of relaxation and recovery. And laughter is contagious, visually through mirror neurons and kinesthetically through resonance: when we hear someone's mirthful laughter and see their bodies rhythmically doubling-over in glee, we tend to respond with laughter. Remember those youthful gatherings when one child's spontaneous laugh set off a cascade of other giggles, snickers, and belly-full laughter? Both smiling and laughing are part of social responses that link cognitive and emotional parts of the brain. The right frontal lobe in particular lets you discern when something is funny. Something fundamental changes in communication when we are able to laugh with each other.

Psychoimmunology is a field that has paid particular attention to the therapeutic benefits of laughter. Dr. Lee S. Berk reminds us that this is not new knowledge; the Old Testament of the Bible (fifth century BCE) states, "A merry heart does good like a medicine, but a broken spirit dries the bone."[6] We recognize that when we have a disturbing thought—like remembering an angry conversation—it registers through instant sensations such as heat, tension, and increased heart rate. Similarly, when we engage humor, we can notice positive physical responses, including relaxation and more ease in communication.

The brain orchestrates humor and benefits from the process. Brain waves are essentially synchronized electrical impulses: masses of neurons communicating. *Gamma waves* are the fastest frequency, associated with bursts of insight and also with laughter. According to neuroscience research, gamma frequencies are now associated with *neural synchronization*—nerves tuning into and talking to each other. Dr. Berk elucidates: "That creates an effective brain; gamma waves are the frequency associated with the highest level for cognitive processing, for functioning effectively, and for communicating—the complete opposite of a depressed state."[7] Thus, laughing itself supports learning processes in the brain by facilitating connections. It helps us find an ah-ha moment within a collision of worldviews or comic juxtapositions, linking insight with the ridiculous.

What makes something funny? Because the mind is predictive, when we watch a comic character slip on a banana peel, we anticipate the fall. We have experienced the sensation! The mind also tends to dwell on the negative, keeping us alert to danger and threat for survival. Since we have more emotions tuned to threat than to joy, we have inherent empathy for dangerous or pathetic situations. We see cartoon characters running toward a cliff and *know* it is dangerous. Or we watch an uncoordinated, socially awkward TV actor bumble through a scene and laugh at the irony of the tragi-comic combination of humor with pathos.

Response to physical, embodied humor varies; to some it is just not funny to get bashed in the head, fall off a cliff, or be mired in hard times. Comedians and researchers offer an overview of various aspects of humor: incongruity (when things don't match as expected), superiority (puts you above another person or group), quick turns of fate (thwarting predictive processes), relief (it's not happening to you), opposites and juxtapositions (combining the hilarious with the disturbing, creating an emotional roller-coaster), puns and questions with a twist. ("Why did the farmer win an award? Because he was outstanding in his field.")

Laughing has such an important impact on our lives and well-being, but what happens when it is stifled? Laughter and its freedom-making partners, dance and music, are often grouped together as suspect and sometimes forbidden because they are powerful. They create a sense of individual freedom and inner strength. When a group is being subordinated to authority, any forms of cultural cohesion like laughter, dancing, music, spirituality, art, and artifacts are often repressed as threats to social order. In school many are taught: "Don't fool around in class; get serious." Humor is restricted in specific contexts as we mature, but we seek it in other ways, like comedies on TV, films, or joking around on our own. Laughter can cut through and offer fresh perspectives on entrenched ideas and beliefs. It can also be a way to cope with tough, painful situations. With enough confidence and practice, we can sustain the ability to laugh with others and at ourselves throughout our lives and in repressive contexts.

When is humor appropriate? Some humor makes us uncomfortable for good reason. Although there is relief in being spared the insult of the comic's attention or the publicized cartoon, humor that is dismissive of social groups, puts down others, and undermines integrity is not to be tolerated. It programs the nervous system into defensive patterns that can take years to undo. The privilege to poke fun at a marginalized group is not ethically appropriate. Bullying that lets someone get away with demeaning comments, short-cut slurs, or derogatory labels because it is delivered in a joking manner is mean, disturbingly impactful, and all-too-common in political and social spheres. While such base humor is not funny, making impactful commentary through comedy requires wit and critical thought. We are responsible for discerning between well-intended and demeaning humor and choosing responsibly.

CULTURAL CONTEXTS

Similarities exist between humorous tales across cultures and periods. In his pivotal 1985 book *Humor and Laughter: An Anthropological Approach*, Mahadev Apte observes: "Not only does humor occur in all human cultures,

Laughter Yoga

I am taking a class in laughter yoga with dance educator and singer-songwriter Deanna. She is a fine teacher. This day, there are only three of us in the small studio; it is hard to get going. We do exercises, then come together to laugh. Facial cues are hilarious—it is hard to resist. She admits, "It works better in a big group." But, like her generous music (including The Jinxes) and her dancing, laughing together with Deanna can be a generative offering in stressful times. She became certified in laughter yoga to help people find more ease. I appreciate the induction.

Regaining Humor

Reflecting on a challenging experience, my friend Kevin sends me an email about the role of laughter in recovery: "I lost any sense of humor for several months, and any sense of pleasure. This was in 2017–18. I saw you when it was ebbing, the last bits draining out the bottom or the top somehow. Then, I remember the day in which someone said something so useful to my consciousness, so bluntly obvious that it burst my conviction of catastrophe, that I laughed strongly, and it became clear how important that humor function is. My sense of pleasure was clearly a vital part of health and had gone AWOL and maybe I didn't know it was gone! I wouldn't have been able to articulate the lack of it. I think when humor is absent, and you see people laughing, it's a strange void, a dull noting of flatness."

it also pervades all aspects of human behavior, thinking, and sociocultural reality; it occurs in an infinite variety of forms and uses varied modalities."[8] The empathetic quality of humor endows it with powerful potential to affect and influence. Even though there are distinct differences in the way cultures and periods create and respond to humor, there are also many overlaps. Tricksters and naïve characters seem to be universal. The Roman *Philogelos*, or *The Laughter Lover*, from the fourth or fifth century CE, is presumed to be the oldest surviving book of jokes. Among its two hundred and forty gags, many are recognizable in modern slapstick or standup comedy. Here's one: "An intellectual came to check in on a friend who was seriously ill. When the man's wife said that he had 'departed,' the intellectual replied: 'When he arrives back, will you tell him that I stopped by.'"[9]

Psychologically, we understand that humor is interpretive and culturally encoded. What is funny to one person or in one cultural context differs with other individuals and communities. Cross-culturally, jokes are challenging to understand: their nuance is hard to translate. Even for second-language speakers who have spent many years in a new linguistic context, the nuance of language makes comprehension difficult. It can make you feel "outside" the group, intentionally or circumstantially. Sometimes you feel yourself on the other end of a joke or invective comment that is perceived as funny by others: you feel shame, a powerful emotion that changes joy and excitement into lingering pain. Humor does require a fearlessness, a willingness to be seen and heard and, possibly, to be misunderstood.

Humor is a state of mind. To enhance well-being in living and communicating, you can pursue humor and laughter as essential refinements in your day. Like wishing to be physically fit and just thinking about an exercise plan, you have to actually engage happiness through laughter and attention to effect change. When a serious message is delivered through the medium of respectful humor, people are more likely to listen, to discuss it with others, and to engage critical conversations.

Connections

Today, you can invite the lens of humor and laughter into your encounters. Laugh three times today, even if for no reason. Or dial into something funny that you know makes you laugh. Notice how this impacts your mood, your communication with others. Humor can reduce the stress and emotional exhaustion that accompanies seemingly insurmountable problems (personal and global) that are part of daily experience. It changes internal chemistry as well as tension levels in a room. Vitality and inclusivity come when humor hits an inner chord, a sense of truth and shared experience between you and those around you. Sometimes laughing is the best you can do within yourself and with others to encourage healing and connection.

TO DO

INVITING LEVITY—FLYING (QIGONG)

5 minutes

Lightness lifts your mood. Looking upward, opening the chest to the sky, adding a slight upturn to your lips can elevate your mood— reducing stress and stimulating easeful enjoyment. Flying brings this upturned energy and levity into your day.

Begin in basic standing posture, arms relaxed by your sides:

- Imagine your arms as wings, hands pointing toward the ground. On an in-breath, bring them just a few inches away from the sides of your body, then relax them to your sides on the out-breath.
- Repeat, flying your arms a bit higher on your in-breath, releasing your arms to neutral on the out-breath.
- Now, pump through your legs with each swing of your arms: extending your knees as you lift the arms (flying!), bending the knees as you drop your arms downward.
- Fly five more times, gradually getting higher and bigger with each flight and release.
- Allow a pendular movement in your arms, so they swing with momentum.
- On the fifth "flight," bring the backs of your hands together overhead and pause. Take a few breaths, then slowly release them down to your sides.
- Enjoy the after-flush of flying, noticing the sensations you have stirred up in your body.
- Invite an authentic smile, cueing ease.

TO WRITE

REMEMBER THE TIME . . .

10 minutes

Some memories are hilarious now, but they were embarrassing at the time they occurred. Tell a funny story; something from your past that you consider to be funny. Let yourself reinhabit the moment with all of your senses alert. "I remember the time . . ." When you read aloud, try accenting different words in each sentence, noticing how accentuation affects rhythm as well as meaning in delivery.

Clowning Around

Writing about humor, graduate student April Danyluk tells of her decision to attend clown school. "As an April Fools' baby, perhaps I was living up to my destiny; or maybe I wasn't fully ready to grow up. In either case, at twenty-four years old I made my way to the Clown Conservatory of San Francisco. The funny thing is that now, ten years and two master's degrees later, I still feel that the skills I gained in the year-long clown program have been more relevant to my life than almost any other experience. Knowing how to engage and respond to an audience, as well as how to turn my weaknesses into 'golden opportunities,' continues to inform my daily interactions."

Dancer (2020), marionette
by Kristen Kagan Yee
Low-fire clay and glaze

Day 19

Resistance and Relationship
With Chris Aiken, dance artist and educator

You learn to become aware of the other person's power in the room . . . And, rather than mirroring, to make choices. — Barbara Dilley, from "The Unravelling of a Dancer" in the New Yorker

Chris Aiken
Photograph by Jonathan Hsu

Resistance to change is part of being human. Sometimes it is experienced as an immovable wall inside us rather than a bridge or an opening to a new possibility. Our hard-won identities—personal and professional—resist external pressures or invitations. Our healthy ego, our "sense of ourselves," deep down inside wants safety rather than change. Survival, fundamentally, is the goal. Social or personal change can evoke fear, and if you dig a bit further, the underlying resistance is about death or loss. Some part of ourselves dissolves and releases its hold as we move into new terrain, new relationships. Will we survive? Will we lose our identity? Will our stability be impacted if we open beyond the known and surrender? There is a choice whether to change or hold on to familiar ways.

Most of us are already in resistance: there is a residue of tension in our bodies from the demands of our days. The overflow of information,

tasks, and multiple modes of communication create strain. The question becomes: How do we use this blocked, tense energy to enhance creative communication and relationships? To move things forward, water can again be a useful metaphor: the frozen, ice-like places in ourselves can melt from warmth into flowing water. Most resistance comes from our mind projecting our expectations into the future or being mired in memories embedded in the past. To invite new mental images and imaginings, we can first take stock of sensations in the present moment. The feeling of resistance can change into a doorway toward fresh response, replacing tension with flow. Noticing that you are okay in the present moment (no immediate danger present) can support relaxation.

Language makes a difference. Sometimes it is words that create resistance. If you change how you frame an experience or a new project, you may receive a totally different response. It might be the timing: what you are proposing is too much too fast. It could be about scale: excessive energy or money will be expended. You don't have to give up immediately when you encounter resistance in yourself or others. There may be ways to flow around, with, or through it: try a different tactic, shift directions. What hasn't been considered? Where can you embody a "Why not?" attitude? Sometimes resistance informs situations in a nonrational way. What seems negative might be positive, offering possibilities that you have not yet discovered. Developing a relationship with resistance can involve staying in dialogue with parts of yourself internally (emotionally) that are confident about not knowing. We can also recognize when to let go and move on, directing our energy in a more useful way.

RELATIONSHIP

Relationship involves both saying "no" and saying "yes." "No" is essential in response to threat or inappropriate actions or words. We need to know we can say "no" to protect ourselves, to deflect the negative impact of other people's behaviors on our sense of well-being. But if there is no actual threat, we can make effective choices to move forward rather than shutting out people or new experiences. Inviting openness, we can stay with the image of flow. Water, although fluid, is a powerful force that can cut through resistance. The Colorado River is a stunning example; the flowing river gradually wore away solid rock, creating the mile-deep Grand Canyon over millions of years. "Yes" is a powerful agent to move things forward, allowing change. Practicing the "no" and "yes" of life can support responsive ease: you know you can take care of yourself as well as open to new situations.

Merging—losing oneself in relationships—creates problems. Although reciprocity and empathy with others are essential, self-awareness is also required. Boundaries are useful. Maintaining a sense of self, of standing on your own two feet when deeply engaging with another, lets you both feel more free. Intimacy offers special connections and challenges; how we speak to each other can inform how we support each other's privacy and each other's growth. And independence can enhance mutuality. Providing

Words
Deciding about retirement after thirty-four years of college teaching, I long for then resist the process. In New Zealand they call retirement "going redundant." Emeritus professor of English John Elder partnered the word retirement with "relinquishment." What is my view? On a self-retreat in Connemara, Ireland, I lie belly-to-earth for hours on the beach in unseasonably sunny weather. In that deeply secure, relaxed state, new language emerges. "Retirement" is replaced with "returning to full-time art-making." Changing the wording opens possibilities for response: I sign the papers.

Interrupting Stuckness
Feeling stiff mentally and physically after hours at my desk, I get up to stretch. Including fascia in my mental image helps—it's like pulling a strand of cloth: the whole fabric responds. I know that fascia is a body system complicit with "stuckness." When we don't move an area, this connective tissue that surrounds every bone, muscle fiber, and blood vessel can get "glued" together. We feel resistance. Stretching my arms in the air and yawning, I feel a return of responsiveness and mobility. A friend suggests: set your timer for every 30 minutes when you are sitting; stand and move to refresh! In a long lecture or online meeting, include "movement snacks" to revitalize your energy. Author Don Johnson describes this as "interrupting the trance of sitting."

Inhabiting the Edge

I talk about change but don't really like to do it. Especially when searching for creative material, what's below the surface is generally where I'm headed. There are reasons that theme and those memories have been pushed down and out of my conscious awareness. So resistance surfaces as a partner to meeting and greeting. Working with colleague Peter Schmitz as rehearsal director for my solos, he intuitively finds the edge where I don't want to go. Usually, it's about being more free, but I have carefully taught myself restrictions: patterns of propriety which were successful in my Midwest-girl heritage. Now, they hold on for dear life!

Practicing "No"

Colleague Chris Aiken, while teaching contact improvisation, offers the following advice about "No": "In my experience with dance, learning to say *No* allows you to say *Yes* to deeper communication. I'm thinking of contact improvisation, where if someone does something that I'm not comfortable with, I don't have to shut down, I can refine the dance by steering it away from what feels scary, unsafe, or uninteresting. This doesn't always work, but for those who are consistently in situations where they feel they can't communicate—either because of insecurity, lack of power, or fear—this can be a way of practicing saying 'No, but . . .' This is the converse of the improvisation strategy in theatre of practicing saying 'Yes, and . . .'"

support requires maintaining your individual spirit and strength so you have something to offer in exchange. Draining all your energy into someone else or into a project is not sustainable. Sometimes you need to spend time alone. There is a descent required into one's own life that cannot be accompanied by a partner; we each have our own life journey. Self-awareness is a generous component of communication that addresses the questions, "Who am I, and what am I trying to convey?"

Resistance to speaking up in relationships or in public settings can be embedded from past experience. Fear is deep-rooted energy; speaking to or in front of people, for some, seems worse than death. Prolonged silence can feel like strangulation, as words cannot find their way past the throat. Speaking is a sensory-motor act; we feel ourselves speaking and hear the response. Context informs: What is the implicated background information in a conversation? How much security and reciprocity are present? Sometimes we have to say something out loud to hear what we are thinking. Speaking, finding a voice, can be a way of detoxifying stagnant, repressed energies. In a group context, David Bohm in his book *On Dialogue* encourages the distinction between discussion (offering your pre-formed views) and dialogue (starting from a neutral place responsive to what is being spoken).[1]

SUPPORTING OTHERS

Although change is a compelling idea, the experience can be challenging. Change has to be self-motivated; there must be a commitment to the process. Change is not packaged in a "one, two, three" learning kit; it is an inner experience, expressed outwardly. As we move past resistance, we may experience disorientation. Moving between the familiar and the new, sustaining ambiguity, requires pause; it is not always comfortable. Once we have taken the journey, we can lead others through the terrain of resistance and disorientation. We should not ask other people to change unless we understand what change feels like in ourselves. Inequity thrives on asking other people to do what we cannot or do not want to do personally. Change can become easier or harder with age and experience: we can become fixed in our ways or greet resistance like a friend. Stepping up to the plate, we move past hesitation into expression.

Perceptual skills support resilience and relationship. On a neurodiversity level, each of us perceives the other differently. How we remain connected and attuned to one another differs. Each person is in a varying state of receptivity and capacity for comprehension in any given moment based on their history and context. If someone misses key information, they are at a major disadvantage—they do not have the information needed to change or adjust in conversation. For some people who have conditions and histories that limit perceptual skills or capacities, or who are working in an intercultural or international context where cues are easily misinterpreted, it can be helpful to have forms, structures, and strategies to support group communication and confidence.

A facilitator may offer practices to ensure everyone has a voice in group discussions. Skillful attention and intentional guidance can reduce conversation anxiety. Inviting one comment to lead to another encourages good listening skills. Tracking who has not spoken considers proportions of time allotted. Calibrating or stacking comments, through hand raising or passing a designated object to signify a change of voice, can encourage succinctness while also welcoming and acknowledging the role of a half-formed thoughts. Individually, as a member of a group, you can learn to relax your gaze, calm your voice, and pause to make space for what might unfold in a lively exchange. It can be helpful to encourage spaciousness and diverse thinking by opening your focus to your surroundings. Broad attention helps us pick up other cues we might be missing, making room for others rather than dominating the scene.

Maintaining your confidence as facilitator or participant can be supported in basic ways. Choosing where or how to sit or stand in a group cues easeful alignment and lets you co-align with others. Broad peripheral focus allows you to take in more of the environment around you and encourages resonance with other people and the place. Spreading the palms of your hands on the surface of a table or in your lap releases your "grip" on outcomes, allowing receptivity to others' views. Wearing comfortable and appropriate clothing offers visual identity signifiers to others and oneself and supports clear focus. An easeful stance creates stability and mobility. Context determines how much leeway you have for creative expression. If you are too preoccupied with yourself, you miss what is happening right in front of your eyes—who is sitting next to you or the synchronicity in relationships or conversations. Opening beyond yourself, holding a big view, relaxing, and tuning to the larger landscape of the moment lets you be inspirable—open to the unusual.

Artists are familiar with resistance and its partner, ambivalence. Resistance to a blockage creates specific sensations in the body; tension, heat, rigidity in the spine. You feel it in yourself or sense it in colleagues or audiences. Resistance is natural but it uses up energy. Motivation is needed, something that gets communication going. Ambivalence is a drain; there is not enough focus to support clear action. One aspect of training in the arts is that you don't let resistance stop you. You recognize it as a possible entranceway to discovery, indicating that you are likely close to something new. With practice, you can give yourself enough support to take the next step without overwhelm or shutdown. This involves the body: finding ground, space, and breath to orient to the challenges and invitations of the moment.

Communicators today must be integrators. We have to not get stuck in one way of thinking. If we know who we are deep in our bones, we can relax and cultivate the flow of easeful interaction. It is a paradox: we develop our views and the specific skills of our field, and then we hold an openness for multiple perspectives. We can look for connections and relationships rather than separations; doorways rather than walls; what is working, rather than problems. Resistance fades with effective embodied communication.

Holding On

Tension in the hips can transfer (through connective tissues) to the tongue and jaw. You have to let weight transfer down to the ground to circulate back up for effective breathing and speaking. There is bi-directionality at work throughout the body. The jaw is *not* a clamp, it is a free, expressive opening. I've often told the story of the first theatre director who tried to relax my jaw by taking it in her hands to jiggle and loosen my hold. I reflexively shoved her so hard she hit the wall behind her. It was the first time I realized I had tension in my jaw. We all hold resistance somewhere; one place of tension impacts the whole.

Eeva-Maria Mutka and Fabiano Culora
London, England (2015)
*Body and Earth: Seven Web-Based
Somatic Excursions*
Still image from videography
by Scotty Hardwig

Connections

The invitation for today is to notice when you feel resistance to a person or idea. What are the sensations that arise in your body, and how do you respond? Sometimes resistance may manifest in ways that you might not yet realize are resistant. Signposts can be tension, heat, avoidance. Resistance is a call to pause, allowing more absorption and conscious attention to the people and events involved. Sometimes just refining the ways you express your ideas or your approach can open doorways to creative conversation with others with different opinions. Explore possible responses to bring more ease and permeability. Resistance can be a friend, revealing your level of investment; it shows you care deeply. Although your conversations, projects, or plans will often meet opposition or lethargy, your relationship to that resistance can change. ◉

TO DO

FACING A WALL

10 to 20 minutes

Feeling the actual resistance of a wall can help release pent-up energy. It gives you something to push against, allowing energy to be released from held muscle tension and recycled back into your system. Sometimes explorations help you complete unfinished impulses to push or shove away that weren't able to be expressed in the actual past situation.

- Find a sturdy wall. Take time to press both hands against its surface.
- Lean your body weight into the wall; feel both its support and its resistance.
- Close your eyes and explore pushing, sliding, leaning, resting.
- Let your body tell you what it needs in this exploration; give yourself time to be "bored," dropping below willful intention to receive the benefits of both engaging and expressing resistance and relationship.

TO WRITE

TALKING SHOES (FROM DOUG ANDERSON)

15 minutes

Our shoes have taken us down many paths, some more memorable than others. Write about one of your shoes (3 minutes). Then create a dialogue between that shoe and another one of your shoes. Let them carry on a lively conversation about their roles in your life (10 minutes)! Push the boundaries of your imagination, giving voice to your shoes (10 minutes). Read aloud to yourself and your shoes!

Digging

Writing for too many hours, I must go outside. Drawn to a nearby boulder viewed from my studio, I begin pulling at thick roots that lift soil from the perimeter. Revealing more rock below and to the side feels related to writing: unearthing bits and pieces. One exposed bit leads to the next. I find an old metal horseshoe, a silver spoon, a giant beetle at home in this soil. The rock is a giant glacial erratic left behind from the ice age, covered through the years. Uncovering is a slow process; I will return many times, scraping, pulling, exposing what was right there all along.

Day 20

Sensations and Emotions
With Mary Abrams, founder of Moving Body Resources

Emotions are a combination of biology and memory at play in the scenes of our lives. — Mary Abrams, inspired by Silvan S. Tomkins's Affect, Imagery, Consciousness

They are real, those moments of activation related to emotion. At a survival level, *sensations* are foundational. Every living cell has three basic characteristics: the capacities to reproduce, metabolize, and respond to changes in the environment. The last, the capacity to respond, reflects the underpinnings of communication. From a well-being perspective, the motivation mechanisms within a cell to move toward or away, to ingest or not ingest, are expressed in much more complex ways through the emotional systems of the multi-celled human being. Awareness and articulation of our ongoing sensory landscape, using descriptive language to identify sensations, makes helpful space between sensations and their sometimes-volatile associations with emotions.

Four terms help us to understand the complexity of this process: *sensation*, *affect*, *feeling*, and *emotion*. Although these terms are routinely used interchangeably, it can be useful to differentiate them to deepen our understanding of embodiment. *Sensation* involves ongoing sensory activations through billions of receptors that inform us instantaneously and continuously of inner and outer occurrences. *Affect* is the body's biological response to these sensations, part of our endowed makeup and occurring below the level of consciousness. *Feeling* is what we experience when affect

Grand Rounds (2017)
Choreography by Tamar Rogoff
Photograph © Harvey Wang

ripples through our bodies and when we become aware of ourselves having a personal and subjective experience that we call a feeling. Affect and feeling become *emotion* as they combine with the biography of our experience and can be named and identified.[1]

Every emotion has accompanying sensations that create an ongoing feedback-feedforward loop between these various dimensions that shape our lives and life choices. As we heighten awareness of sensation, we can "listen in" to our body's internal responses to mental processes and become more conscious of what sensations, affects, and feelings are at play within our emotional experience. In embodied communication, valuing sensations as part of the experience of emotions can support awareness of reactive responses, clarifying exchanges.

SENSATIONS

The sensations related to emotions can lead to mood and to meaning. In humans, complementing the trillions of responsive cell membranes that comprise our bodies, specific sensory organs are located throughout our tissues, along with the special senses located in the head. Together, they combine to form an interconnected communication system throughout the body-mind. Downgraded as "second cousins" to thoughts in some cultures and communities—like many schools and workplaces—the sensations that define our lives inform our bodies in ways we might not notice. For example, the heart-felt experience of a place, although hard to describe, underpins our capacity to be present and available in the moment.[2] Our resonance with a person, or situation, or the intensity of an exchange draws us toward or away. The "silent-level processes" underlying our actions and interactions inform everything we do.[3]

AFFECTS

As the body's way of preparing for action, *affects* modulate the intensity of response based on circumstance. Understanding this pre-feeling, nonverbal baseline gives agency when working with our sometimes-confusing emotional interactions. Silvan Tomkins (1911–1991), in his magnum opus *Affect, Imagery, Consciousness* (which was published in four volumes over thirty years), describes the roots of how we develop emotionally by teasing apart biology and psychology. In his view, there are positive, neutral, and negative affects that move you toward or away from a stimulus. Mary Abrams, director of Moving Body Resources, explains: "Emotions are scripts that we develop, mostly unconsciously, as we experience affect and feeling in the scenes of our lives, alone and with others." Affects are messages about our well-being that focus our attention. Thresholds of stimulation are based on personal experiences within familial and cultural norms.

According to Tomkins and ongoing colleagues, there are nine inherent affects that we can explore. Each exists on a continuum from lower to higher intensity. We can change the threshold of escalation for ourselves, impacting the density of stimulation and our reactivity. 1) Interest moves to excitement. 2) Enjoyment to joy. 3) Surprise to startle. 4) Fear to terror.

Re-Opening

Visiting our massage therapist in Maine for the first time in the months following Covid-19 regulations, I welcome her skilled presence. We wear our face masks, and I settle onto the massage table, closing my eyes, relaxing. I am always mildly interested in what thoughts float to the surface when she works on different areas of my body. As she brings her clear touch to my right hand and forearm, I recognize I am still holding onto my friend's death. Grieving comes in layers. As she stretches my palm, massages out each finger, circles my wrist, I try not to analyze, just stay with the quiet release of tears.

When it's time to turn onto my belly so she can work on my back, I realize (crunch) I'm still wearing my sunglasses! We have been so "in disguise" in chilly spring weather (my black Icelandic fleece on sixty-six-degree days; sunglasses, face mask, and hat when we go out; a full-net bug shirt to protect from black flies) that I was unaware. *Surprise* doesn't escalate to *shame*, just humor. I laugh: "Have you ever given a massage to someone wearing sunglasses?" Her response: "No—I thought you were protecting your eyes!"

When we finish, I walk outside to stroll by the river and enjoy the hot sun. Relief—there is no hip pain and there are no bugs! Stripping down layer by layer to reveal bare-skinned arms and a bright-colored, open-necked shirt, accompanied by relaxed muscles, I realize there is a process involved in "re-opening." The first step is meeting each other once again, skin to skin.

Surprise! Ewan Elijah discovers his feet,
age five months (2020)
Photograph by Whitney Will

5) Distress to anguish. 6) Anger to rage. 7–8) Dis-smell and Disgust to distancing. 9) Shame to deflation. We see these affects at play most clearly in babies as they learn to navigate communicative terrain. Surprise can escalate to startle, resulting in a fit of crying or laughter depending on the degree of perceived risk or threat. Noticing the ways that we habitually respond to inherent affects is a way to orient to the biology of emotion as we resonate with others. We are "affectively resonate" all the time. Some of us have higher thresholds for certain kinds of situations, impacting intensity levels.

FEELINGS AND EMOTION

Feelings comprise our *felt sense*—sensations linked to interpretation and naming. Feelings are personal, part of our psychological matrix. Feelings reflect our values—what we care deeply about, stand behind, and might defend with our life and livelihood. They involve evaluation and judgment of what something or someone is worth and are influenced by past experience and reflection as well as anticipation of future consequence. *Emotion* is the display of feelings—private or public, authentic or contrived—that is sometimes disguised for appropriate social engagement. Emotions impact others and spread into the surrounding space, sometimes eliciting triggered responses, shaping communication outcome.

Emotion implies movement. Allowing emotions their expressive flow can feel like riding a wave. If you look at a photograph that creates a strong emotion or remember a personally charged encounter, you will likely notice the rise of sensations in your body. Let your response ride on a breath into a word or sound like "Ahh!" If you follow the waves of emotions rather than trying to control them, you can become familiar with the gathering of energy, cresting, surging forth into expression and the dissolving and returning to the oceanic matrix of the body. Each phase is full of sensation, energy, vitality, and risk.[4]

As a powerful stimulator of hormones, emotions influence key body processes—the bowels, blood sugar, and heart are particularly impacted. Emotional stress can change metabolism in ways that are hard to trace and treat. The immune system is notably impacted; stress reduces immune functioning as the body puts its resources toward the perceived threat rather than general immunity. Returning to homeostasis involves working with the organs as key players in vitality and health. A deep breath, a yawn, or momentary pause after an in-breath can help reset the autonomic nervous system that affects all our systems. Neurologically, we know that the brain can change itself; neuroplasticity is part of brain health.[5] Activating new neurological connections can refresh emotional states in support of resilience.

Although certain emotions are basic to survival, our attitudes about those emotions can and do change throughout our lives, affected by our maturation and life experience. Something you might once have been afraid of, like auditioning for a play or facing confrontation, can change with experience. Emotions reflect what you have been taught is normal and what you value. Yelling loudly in some families is a sign of love and involvement; for others it represents aggression. As a form of energy, emotions can be channeled in different ways, toward longevity or toward peak performance, for example. Noticing how you expend your energy lets you monitor the ways your body is impacted.

Mood is prolonged emotion, a feeling state that pervades the body as the endocrine secretions move through the circulatory system. That is why movement can change our moods; it facilitates the pumping and processing of hormones and stimulates a new set of directives. Mood can be the pre-effect or the afterwash of an experience: a holding back of a wave of emotion or the resulting calm after hysterical laughter, an angry fight, or a good cry. Emotions, like joy and irritation, are not just thoughts about joy or irritation. They are experiences; they are not abstractions of experience. Thus, it is difficult to separate an emotion from identity. The usual expression is "I am happy," not "I am thinking happy thoughts." Yet attention to sensation helps us move through and with changing states, recognizing that self-awareness is an essential partner in all aspects of experience.

REFRESHING RELATIONSHIP

The sensations that accompany pleasure can need reinforcing, tuning up, and freeing. The frequency of responses like interest or enjoyment can be cultivated and developed through attentiveness. How often do we laugh, smile, and allow whole-body joy? Release is not just about doing. Sometimes we "think" we are enjoying something like a walk on the beach, conversation with a friend, or dinner out on the town, but actually we are distracted—busy thoughts just keep dominating the scene. Allowing pleasure to move through us and offer nourishment is another dimension. Attitudes and distractions that limit joyful interactions may need adjusting. Responsivity can involve unhooking learned dampers on internal delight.

Somatization is a word that reflects awareness of embodied experience. It can have positive or negative connotations depending on context. Psycho-somatic illnesses can be trivialized because they link emotionality with physical symptoms. Yet we all somatize our feelings; every thought has an accompanying sensation. Somatic symptoms are real and help focus attention and diagnose illnesses that may be hard to locate. Embodied *somatization explorations* can calm stress and deepen self-understanding, enhancing the relationship to the body. Basic to the process of discharging stress, sensory scanning can offer a way out of repetitive, limited, and sometimes harmful patterns of reactivity linked to past trauma or conditioning. Opening the lens to sensation enhances awareness of the present moment and offers choice.

In his video about "The Art of Noticing," Dr. Peter Levine describes how focusing on sensations can be useful in healing trauma. His global training programs in Somatic Experiencing offer a systematic process with therapeutic goals. "Helping clients make the connection between their bodily sensations and a traumatic experience often helps them view these triggered responses in a new light and eventually discover new ways of

Multidimensional Range

My teacher in graduate school, Dr. John M. Wilson, combined dance technique and philosophy. As we practiced our challenging turns, balances, and complex rhythmic sequences, he would speak about the multidimensionality of the human spirit. We learned and embodied the knowledge that every human being has a full range of expressive capabilities from detailed gesture to raucous dancing. It was our work to construct how to utilize this range in our creative lives and personal presence in the world. His three impactful courses in human anatomy, dance technique, and aesthetics and philosophy intermingled during the days and nights of our college education and traveled with us into our professional careers.

Lupine Walk

Early morning light invites a lupine walk along the stream. The cone-shaped purple and pink flowers on their long sturdy stalks are all opening; when the petals fill out at the top, the short-lived blooming-period passes. They line the edge zone of the path, preferring poor soil, full sun, sufficient moisture. I'm told our minds turn more frequently and quickly toward negative thoughts than positive—noticing threats for survival. I drink in the beauty, taking small sips, leaning into joy. There is a flash of guilt, filling up with pleasure when so many are suffering with racial injustice, illness, death. Perhaps a dose of joy is needed wherever we find it, for a few moments each day, to keep moving toward life as it could be.

healing." Moving from overwhelm to empowerment can be supported by skillfully differentiating sensations linked to a past experience from the sensations actually happening in the present moment: "moving feeling out of the body."[6]

Orienting to the presence of sensations, affects, feelings and emotions in your communicative interactions can support and help to refine easeful engagement. Somatic practices address each of these embodied states; you utilize what you understand and value. Attention brings agency between activation and how you choose to respond.

Connections

The invitation today is to focus your attention on two affects and how they manifest in your encounters. Specifically notice *interest* and the possibility that it increases into excitement during your activities. Catch moments of *enjoyment* that might just escalate into joy. Remember that you experience these biological aspects of emotions spontaneously and on a continuum of reactivity, depending on the degree and length of activation. Observe how *interest* and *enjoyment* register as *sensations* and interconnect with *feelings* based on your personal history and social context. Today's exploration is a first step in opening awareness to the full span of affects in your living and communicating. Their impact on emotional complexity and authenticity is a foundational aspect of your endowed human nature.

TO DO

FIERCE TIGER (QIGONG)

3 minutes

Activating sensations and emotions through the face and hands can be fun.

In firm, wide-leg stance, knees bent:

- Bring your hands in front of your shoulders. Add tiger claws to your fingers.
- Bend your knees further in a deep strong stance.
- Activate your eyes, stick your tongue out as far as you can—look fierce!
- Pause and recover.
- Repeat: Deepen the bend in the knees. Exaggerate your facial expression. Powerful stance, powerful claws.
- Repeat a third time: Start with a deep in-breath; breathe out with a growl as you bend your knees and engage the full fierceness of your pose. (Imagine a snapshot of yourself in your expressive pose—a guardian figure!)

- Relax and notice what you have stirred up by embodying your Fierce Tiger, warding off negativity. You have likely stimulated "surprise" and all the layers leading to emotions, including sensations, affects, feelings, and resulting emotions. Hopefully, it was fun!

NOTE: If you can't follow the cues, make up your own version, finding the tiger in you!

TO WRITE
E-MOTION
10 minutes

E-Motion implies movement. Write an anecdote about a time when you felt strong emotions. (An anecdote is a brief revealing story about an incident with a beginning, middle, and end. Often it is a story you have told more than once because of its humor, impact, or insight—like the stories in the margins of this book.) Allow your conscious mind to take a back seat to emotional expressivity. Then write stories from ages six, twelve, twenty . . . how did your relationship to emotions change?

Freedom

The inner critic is powerful for many of us: an inner voice commenting on our every move and behavior. When Mother was in a nursing home in the last years of her life, we would give parties for some of the residents to try to cheer things up, break up the monotony. We brought paper hats, cake, and party favors, including blowout noisemakers in multiple colors. Around the table, one usually silent woman was beaming, repeatedly blowing her noise-maker at those across the table. She exclaimed, "If my mother could see me now!" That inner mother's voice had shut down her behavior all these years. She added, "This is the most fun I've ever had."

V
EXPERIENCING INTER-CONNECTIVITY

Barong Ket (2011), mask
Ida Bagus Anom Suryawan, artist
from the village of Mas, Bali, Indonesia
Wood, paint, hide, human hair,
metal bells, gold leaf, silver, glass beads
From the exhibit *Sacred Realm: Blessings
& Good Fortune Across Asia*, Museum of
International Folk Art, IFAF Collection,
FA.2011.44.1[1]

Vision and Insight

Vision is how we see. It is also how we imagine others, the world, and the future. Seeing is a learned experience that develops and changes throughout our lives. In communication, no two people see the same thing, due in part to sensory preferences and personal interpretations. The health of our eyes and the state of our attention play important roles—are we enjoying our day, emotionally distracted, over-caffeinated, or simply alert? Sometimes we limit our vision in communication by holding on to ideas and expectations. Eyes change, and so do we.

Vision is both inner and outer. Opening our eyes involves a dialogue with light we call seeing. Recognizing the emotional significance of light, we feel the impact of what is illuminated or obscured in our visual field. Closing our eyes or softening our outer focus invites us to become aware of the interior workings of our body-mind. This process of inner-seeing encourages questions and discoveries. Sometimes insight emerges in meditative practice, in nature, or in unexpected moments. Imagination plays a cognitive role by connecting inner and outer vision. We construct our view of the world by how and what we see.

Omnidirectional vision includes spherical space: the whole body and the full space around us are included in our visual field. Although our eyes are forward-facing, our mobile neck and spine allow us to turn easily toward stimuli, and all surfaces of the skin have light-sensitive sensory receptors. At any moment we can focus attention on specifics while remaining open to the larger picture, circulating our attention to multiple aspects of the tangible world. Newness presents a challenge: if we have not seen something before, it may not register at all! Once someone points out a whale spout or an unusual feature in a new landscape, we can spot it, literally "making sense" of those shapes and forms. Understanding our eyes, recognizing visual habits, and noticing cultural contexts refreshes communication choices as we enjoy the remarkable ability to see and respond.

FACTS

Seeing is a process. Light sources bounce around inside our eyeballs, triggering neuronal responses in the *retina*. Composed of hundreds of millions of cells, the retina is a thin tissue that lines the inner surface of the back three-quarters of each eyeball. Specific photoreceptors in the retina tell us about shades of light and dark (*rod cells*) and patterns of color (*cone cells*). Cone cells are particularly abundant in the center of the retina, with two million packed into a tiny depression known as the *fovea*. Because of the high density of receptors in the fovea, we have detailed perception of color and form in the center of the visual field. Nerve connections from the

My slogan is "wide-awakeness." To be awake is to take risks, to see things that you probably would not want to see. We have to teach that—an awareness, a courage to see. — Maxine Greene, from an interview

Color and Contrast

Blue. I'm looking at blue. Blue sky, blue water, blue pants, blue boat. Shades of blue are all around me. Some days I like to choose a color and take a walk focusing on everything in that color. My intention makes one color pop out. Seeing is selective, based on intention. It is good to know, to practice our role in the process.

Yesterday I reframed my father's watercolor painting of daisies. In the process, I had to choose a color for the mat that surrounds the picture; it made some colors more vibrant, others fade. Then we added the frame; do you want a similar color or contrast? I realize it's not just what's in the painting; it's the juxtaposition of all the parts. Every added dimension alters the whole.

rods and cones convert information into electrical impulses for the brain. A *blind spot* is created where these neural circuits converge to form the optic nerve, creating the only place on a healthy retina that can't respond to light.

By some estimates, as much as 50 percent of the neural activity in the brain is involved in processing visual stimuli. This dedication of resources reflects the evolutionary importance of vision for the survival of our species. The eyes can be considered exposed tissue of the brain. In the embryo, they are part of the forebrain and migrate to the exterior of the face in the first trimester.[2] (This frontal location allows us to perceive distance and adjust the overall state of alertness in the body.) The *optic nerve* of each eye in humans travels all the way to the *occipital lobes* in the back of the brain, the primary visual center, for preliminary processing. At this juncture with the occipital lobe, conscious recognition requires interaction with other "higher" regions of the brain that are informed by memory, association, and expectation—our body history and worldview.

Seeing is selective. Only one-one-thousandth of the data stimulating the rod cells in the retina goes on for visual processing. What you have seen before gets through visual gateways first. A second selectivity gateway occurs at the *thalamus* in the center of each brain half, which process all incoming impulses from sensory pathways (except smell). At this important juncture, seeing combines with other senses prior to arriving at the occipital lobe for processing and further interpretation: there is no objective seeing. The body almost instantaneously scans for survival threats, sometimes triggering our central stress response system (the hypothalamic-pituitary-adrenal axis). A flood of whole-body physical and psychological actions including mood shifts results. Much of this process occurs faster than we can think, beneath our conscious awareness.

Eyes move to gather information. In fact, they are in constant reflexive motion. These micromovements called *saccades* are so tiny that they are almost imperceptible, adjusting in miniscule stops and starts. Occurring in time frames of about one-tenth of a second, they combine in our brains to form a seamless picture of the world. This process lets us connect the letters on this page into words and sentences while scanning for cues at the edge of our visual field. During the pauses between these oscillations, the eyes process incoming visual data. As the eyes move around the visual field, they allow the small area of the fovea to gather details according to the viewer's intention. Eyes also utilize slower tracking movements to keep the fovea aligned with the intended subject. Vestibulo-ocular movements stabilize and adjust images on the retina in relation to motion of the head; when we look at an object and tilt our head, the eyes compensate to keep the image upright.

Binocular vision offers perception of volume and depth, a three-dimensional view of the world, by giving each eye its own unique view. It also requires entrainment: focusing two eyes individually toward the same image. Six *extrinsic muscles* attached around the soft globe of the eye make this, along with movement in all directions, possible. The soft eyeball, about one inch in diameter, is encased in the *sclera*, a tough but elastic white coat of connective tissue that can easily be deformed by too much muscular tension. Holding or fixation in how one uses their eyes is limiting.

There are also six *intrinsic muscles* within the eyes. They adjust both the size of the pupils in response to changes in light and the shape of the lens in response to depth of focus. The chamber behind the lens is filled with a clear gelatinous material, the *vitreous humor*, that comprises up to 80 percent of the eye's volume and helps to maintain the health and shape of the eye. Each eye is housed in a bony socket (about two inches deep) where seven skull bones meet. Impact to the skull—like a concussion in sports, an accidental fall, or tension in the skull muscles—can affect the bones, causing strain or headaches. That level of distress affects our vision.

Exercising all these eye muscles keeps things flowing by enabling them to adapt to the demands of the moment. Relaxing the eyes can allow something new to be seen and experienced and can conserve mental energy, calming the mind. Palming is a practice where you gently place the palm of each hand over the eyes to offer darkness, warmth, and relief from overwork. Using imagery can also expand your ways of seeing; imagination and perception are linked in the brain. You can visualize eyes all over the body to enhance omnidirectional awareness, one eye looking inward and one eye looking outward to balance inner and outer attention, or a third eye in the forehead as a gateway to insight and inner vision. There are many different ways of experiencing and influencing vision.

Seeing is both the slowest and the fastest sense. Because vision involves complex processing by the brain, it can take time to recognize what you are observing. Yet vision is also the fastest sense, accessing quantities of information in a single glance for survival reflexes. We often fill in visual information, jumping to conclusions or misreading gestures with partial visual cues; this is known as the completion phenomenon. For example, we see two dashes and envision a solid line, or we see a hand reach for a pocket in a combative situation and imagine there is a gun.[3] For speed of response, some visual cueing is essential for survival, but other automatic reactions limit or derail experience. Part of visual training is allowing ourselves to take in more of what is available in the moment so we increase the accuracy of our responses. Especially in intercultural and international contexts, this allows us to interact with, rather than react to, new experiences.

We configure the body both to see and to be seen. Call someone's name, and they turn toward you to see what you want; you look directly at a friend in conversation to connect face-to-face. If someone is viewing you through a camera, it is hard to not arrange and rearrange yourself to be seen. When performing in a film, the instruction may be to not "act"—to be natural in relation to the interconnectivity of what is happening. Yet studies show that observation affects the observed. You also never know where a camera or a viewer is focusing—on your hands, the twitch of your lip, or the person behind you in the room. This is true in conversations as well, and cultural norms vary, influencing how we define an appropriate gaze.

Eyes are impactful communicators. A blank stare somehow tells you

Insight

When my mother had a stroke, we drove the 1,500 miles between Vermont and Florida to arrive at her bedside. I walk in the hospital room, and she looks up from her bed and says: "I just wanted you to be your own original self." Sometimes, change cuts through memory and family history. Truth emerges of its own sweet accord, no barriers.

Direct vision: Scout, Slate, and Maverick *(l to r)*
Photograph by Lysa L. Olsen

that someone has shut down (without telling you why). Some people penetrate you with their stare, fixating on you with their gaze to an uncomfortable degree. Others don't seem to see you at all; their focus is elsewhere as they hold a preconceived view. Reciprocity requires a receptive gaze: taking in, responding, circulating attention. Good communicators practice multiple eye strategies: keeping track of the whole while attending to detail. Constant focusing and refocusing are necessary to be inclusive and open to surprise.

Eye focus occurs on a continuum from a specific to a broad view. *Direct vision* entrains your eyes toward details, pinpointing something of interest in the surrounding space that you want to identify and understand. To experience direct focus, let yourself be drawn toward something of interest and look at it in detail. *Peripheral vision* (also called panoramic vision or optic flow) involves spreading your attention to the edges of the visual field. Although we are often told to make direct eye contact to affirm confidence in communication settings, if we focus just on the eyes, we can miss essential whole-body cues or ignore the context that is informing the exchange. Broadening focus lets us take in both more content and context.

Language around vision guides and records experience. It is prescriptive as well as descriptive, telling us what might happen and reflecting on what occurred. For example, the words "vision," "imagination," "insight," and "envisioning" use the physical processes to evoke and describe something metaphorical, philosophical, or spiritual. Vision and imagination are cortical processes, originating in and utilizing visual centers in the brain; insight and envisioning can be described as arising from the body, utilizing various body systems and cellular processes. We need both brain-down and body-up synthesis for coherence in embodied communication. Our vision of life includes who we are and our unique role in the context of others, community, and the places we inhabit. It is the tether to meaning that keeps us focused and functioning, that makes us open our eyes in the morning, and welcomes the darkness of sleep for dreaming and rest.

Drawing by Avera, age 11

Re-perceiving depth requires conscious attention. Many of us spend multiple hours a day focusing on screens; eight hours is a conservative figure for Americans. Our eyes are entrained to two-dimensional screen space and particular light sources, and our bodies remain unattended. When we look at the actual world around us, we may not be used to activating depth perception in three-dimensional space. Recognizing that substantial visual processing occurs in the back of the brain (occipital lobes), we understand that we are part of what we are seeing, not separate as viewers. We can re-view the world with this intention, seeing ourselves inside an experience—actively engaging visually—rather than separate from it. We can also create appropriate boundaries to counter too much screen time by consistently engaging physical movement practices to strengthen and energize the whole body.

Refreshing visual patterns invites curiosity. For example, when walking through a park with a young child before school, the wonder of the world is fully present. An acorn on the ground is a new discovery: its shape, texture, and little brown cap that can be removed. This involves seeing before naming—before learning that this is an acorn that has fallen from an oak tree and is a favorite food for squirrels, before memory and association layer meaning. Part of adult seeing involves staying open to fresh vision and the potential awe that partners joy and links to inquiry and imagination. This level of seeing keeps us walking down our individual paths and discovering life freshly through our unique eyes, newly amazed, newly humbled at the vastness of perceptual possibilities.

Connections

Today, the focus is on appreciating your amazing eyes. As you flutter your eyelids open each morning, the world presents itself whole and available. If you are living without sightedness, other senses combine to fill in those details. Today, be aware of the ways you utilize your eyes as part of a full-body sensory system in different communication scenarios. Notice where your eyes are drawn and what you avoid. Be aware of the close, middle, and distant ranges of seeing to move beyond the prevalent demand of "screen time." Being conscious of personal patterns lets you enjoy a variety of relationships and communication possibilities that enhance interconnectivity.

where the sky embraced us /
donde el cielo nos abrazó (2018)
Contemporary Dance Company
of the National Ballet of Ecuador
Choreography and videography
by Scotty Hardwig

TO DO

REFRESHING SEEING
10 minutes

The eyes and "I" are often intertwined. In this somatic excursion,
you are invited into fresh ways of seeing for daily practice.

Standing and walking, explore:

- Active seeing: let your eyes lead you in space. (Eyes as
  motivators.)
- Receptive seeing: let your eyes rest, receiving the world around
  you without having to do anything about it. (Eyes as open
  windows.)
- Inner seeing: let your eyes look outward while also connecting
  with your inner, gut-level experience. (The metaphor of one eye
  looking out, one looking in.)
- Place seeing: let your environment or context be more in focus
  than the people around you. (Like a rock climber focused on the
  rock face, an architect on the building.)

- Other seeing: imagine you are looking through someone or something's eyes. (Like a person of a different gender or cultural lens; or an osprey, leopard, or dog with different perceptual capabilities.)

With a partner: Explore this sequence in a group while walking in a room, using your imagination. Meet with a partner and discuss what you found. Which lens felt most familiar? Which was most distant?

TO WRITE
SEEING AND BEING SEEN
10 minutes

When we arrive in a new place, we need to be able to change perceptual lenses to perceive more than our habits. Even when we think we aren't seen, someone is watching. Seeing and being seen both have impact. Find a comfortable place to sit outdoors, undisturbed. Look around and notice where your vision is drawn. Describe this object/place/person in detail. Give yourself time to move beyond first impressions; stay with descriptive (not interpretive) language. Now, switch the lens and visualize that the object/place/person is witnessing you. Describe what you would look like from "their" perspective in detail.

Claiming Your View
Claudio, in William Shakespeare's play *Much Ado About Nothing*, asks: "Are our eyes our own?"

Day 22

Identity and Fluidity
With Nükhet Kardam, professor emerita at the Middlebury Institute of International Studies at Monterey

Becoming a part of "we" does not mean losing a "me." — *Daniel J. Siegel, from* Mindsight

Undivided/Divided (2013)
Shen Wei Dance Arts
Installation at Yerba Buena Center
for the Arts, San Francisco, California
Photograph by Stephanie Berger,
courtesy of the Yerba Buena Center

Identity includes how we perceive ourselves and how we are perceived by others. Our physical body is home ground, the medium through which we experience and connect with the world. Response patterns and communication strategies are implicated in this connection. Three lenses can be useful when exploring this multifaceted topic: personal, professional, and collective identities. Personal identity is our own; we have the ability to observe and refine it. Professional identity includes our work and training; we share it with others. Collective identity involves associations with groups through social constructs such as nationality, ethnicity, religion, race, social class, gender, and sexual orientation. Fluidity—moving between these identities with ease and authenticity rather than rigidity—involves

self-awareness and permeability. Discernment in this process helps us become more effective, self-aware communicators who are tethered to embodied awareness, attentive to context, and open to discovery.

Life conditions will change, and what was once a background aspect of one's identity may become foreground. An injury or illness may change your ability to move or compete as you once did. Professional identity may take a back seat when you have children. Recognition or status as a member of a family or community may disappear when you move or are relocated to another part of the world. Fluidity allows stability in changing circumstances. Change is ever-present in our lives. Sometimes amidst disorientation you feel identity-less: unseen by others or yourself. You may need to consciously shift circumstances to seek the people, place, and work situations that draw out your most authentic self. Fluidity between the multiple identities in your life can help stabilize disorientation as you transition from one period or place to another.

PERSONAL IDENTITY

There are multiple ways of viewing personal identity. It can be useful to name a few and reflect on how these dimensional aspects of ourselves communicate with each other and relate within the overall framework of the body. Distinctions we can consider are the physical, mental, emotional, intuitive, and spiritual aspects of self. However we name the parts, various dimensions of personal identity are in conversation throughout our daily activities. Understanding how much communication and interconnectivity is going on internally within this capsule of selfhood lets us maintain a sense of humor and inquisitiveness about our human complexity and creates the potential to gain clarity. Inner communication can be challenging or easeful; often we listen to one part of ourselves more than another. For example, the physical self and mental self may not agree on the best course of action within a day: you are exhausted, but your thinking mind says keep going. Or you have an intuition about what needs to happen, but your emotional mood keeps you from taking action. There are also notions about a core or essential self that is constant and unchanging throughout our lives. Also referred to as soul, life force, or true nature—with many linguistic variations—this core is considered an integrating aspect of the personal identity we generally name as I, self, and me.

If we are self-aware and embodied, a multiplicity of personal identities can flower and complement each other. We are less likely to give conflicting signals in communication. Congruency, through a sense of our core, lets us be better listeners who are more fully present to hearing and receiving. When aspects of personal identity are restricted or confused by professional limitations or collective views, we may feel alienated from ourselves and others. And when inner confusion around self-identity dominates, it can be hard to focus on the larger picture when communicating with others. Recognizing personal attitudes, assumptions, and fear triggers within the complexity of personal identity can be a first step in clarifying intentions and opening to broader perspectives in our interactions. Our

Email from Nükhet Kardam
as part of our creative conversation:
March 22, 2020. "I was thinking of you as I took my beach walk today. It occurred to me that identity is so important to embodied communication because it can either block communication or open it up depending on how one engages in one's 'identities.' If we get attached to them, we communicate from that place which could restrict us, whereas if we can stand a little apart from those 'roles,' our communication could become 'embodied,' full, open, in shared space. When I say 'core or essential self,' what I mean is to be able to hold those different identities (conflicting many times) a little from ourselves, stand back, and sink into that core self (however we define it) that's deeper than those roles, which not only connect us to our own heart/soul but also to each other because that place is the place of unity. We dip into our Source. As such, our awareness expands, we hear, we see, we touch, we hold. We can 'witness' all of these and yet be present. That's the beautiful paradox."

Naked

Before entering the performance installation *Undivided/Divided* by Chinese choreographer Shen Wei, we audience members
are instructed to leave purses, coats, and
personal belongings at the coat check. This
site-specific interaction is described as
"a vivid fusion of East and West, of Chinese
opera and modern dance," and the packed
audience roams elbow-to-elbow through a
grid of white panels and plexiglass cubes.
Vignettes display dancers, mostly naked,
often with white-painted bodies and pans
of paint, moving individually or in pairs in
relation to an aggressive percussion score
while exposed to our gaze. Vulnerability is
evoked in both audience and performers.
Yielding my driver's license, passport,
credit cards, and outer layers of clothing
is a stripping away: Who am I without
my outer identification? Do I trust that
someone will give it back? Relinquishing
my identity signifiers to strangers in a big
crowd is disquieting, knowing they might
just be difficult to retrieve. It changes how
I experience the whole immersive event.[1]

relationship to our integrated personal identity changes throughout our
lives in differing ages and stages and with unforeseen occurrences. Each
day, "How are you?" takes on new meaning.

Sustaining a coherent personal identity involves making choices.
When decisions need to be made quickly, we respond to sensory information that registers throughout our bodies in heart and breath rate and
gut-level reactivity. Cortical interpretation and reflection are slower, requiring longer processing time, and come in a secondary wave. Sometimes
we need to consciously pause and absorb a situation, listening longer to
the multiple voices within before drawing conclusions, to increase accuracy and assess consequences. Flow in this situation may require firm
identity boundaries and discernment to protect one's own energy from
over-involvement. For example, the banks of a river, the sturdy casing of
a garden hose, and healthy arteries direct fluid flow, preventing leakage
and drainage of resources.

PROFESSIONAL IDENTITY

Professional identity makes us members of a group that is defined by our
specific expertise and work circumstances. There is a role to be enacted
and fulfilled through our work in the world. In a skilled profession, we learn
to adopt the associated terminologies, code languages, norms, and values
which affirm membership. This tends to exclude those in other fields, but
it creates solidarity and ease within a professional community through
social identification and shared interests. When working in international
contexts, professional identity may be central. But if we are limited only
to our professional identity and vocabulary for viewing the world, we miss
other dimensions of engagement.

Cultivating professional identity while remaining open to other perspectives is a layered process. When working collaboratively with colleagues, we may interact with individuals and groups who speak different
professional languages. Likewise, we may also work with those for whom
speaking English (or the designated language for the group) is a secondary
skill. Multiple modes of communication may be essential to clarify communication goals. Nonverbal language has impact: the tone of voice and
the gestures and body postures that convey meaning all come into play.
For example, if you are a doctor who is publishing research findings that
led to an innovative product, you may speak with collaborators abroad,
lawyers at your medical institution, CEOs of business partners, and the
communications department assigned to market your work. International,
interlingual, and interdisciplinary perspectives are all part of the ongoing
conversations.

Values have impact. A more inclusive view of profession acknowledges
that income-producing work may not be the root of one's professional
identity. A job may be a means to an end, the financial stream that allows
us to work at something else we care deeply about that is not financially
productive, like shepherding an environmental project, choreographing

with professional colleagues, or writing a memoir. Family, food production, spirituality, and living an integrated life may be more central to one's professional identity—even if these endeavors are not recognized, valued, or rewarded by society. Caring for children, for example, is serious work that requires skill and perseverance. Cultivating land and food, similarly, is essential and necessitates refined knowledge of natural systems. Feeling useful and productive by working on something that is meaningful to oneself and others feeds personal development, releasing restrictive tension and adding vitality to life.

COLLECTIVE IDENTITIES

Collective identities categorize. Regardless of whether we want to be viewed as part of a group by others, collective identities are part of life. When we encounter groups or engage new communities, collective identities feed our orienting system, helping us form quick distinctions to "make sense" of the sensory stimuli entering our systems. Religious affiliations, political parties, and social groupings (like sports or music fans) reflect collective identities. For survival socially as well as physically, categories and groupings are part of understanding what is happening around us and our relationship to it.

Collective identity can give a person a sense of history, a coherent narrative. If we are part of a social group, we have an extended family and a role to play. It is important to feel we belong somewhere, have links to other people and to the land we inhabit beyond self-orientation and work. There is comfort in recognizing broader connections, knowing there are others with shared values with whom to exchange views, socialize, and expand perspectives. Caring for each other is part of the process and allows us to find support when we can't face grief, illness, or life transitions on our own. Within our appreciation for collective identity, we can notice the delicate weave that holds meaning within a community even amidst change.

Collective identities can also be contradictory, confusing, and even dangerous. Because collective identity quickly signifies who is in a group and who is not, generalizations and misperceptions are inherent. This can limit communication possibilities by determining whether and how to interact based on past experiences and assumptions. Shame or defensiveness around a hidden aspect or misrepresentation of collective identity can foster miscommunication. For example, you may be seen as belonging to a certain racial profile or a social or economic class that is assumed to have qualities that don't reflect your deeper values. Or you may need to defend yourself against stereotypes around gender or culture. Projection also occurs—the tendency to assign our own feelings and fears outside ourselves, onto other people. For example, if we feel threatened in a situation involving someone from a different religion, we may describe that person as threatening. But this comes from our personal experience (the fear response in us) and may not be an accurate reflection of the current context. Focusing on collective identities facilitates quick responses, helps us feel

Being American

Sometimes you don't recognize national identity until you travel abroad. When I studied in Paris in 1969 during the student riots and the Vietnam War, the ugly American tourist was a stereotype to avoid. Traveling in New Zealand during the distressing invasion of Iraq, we found ourselves saying we were Canadian. In China, my nationality was mostly obvious: as I danced and played my snake drum in the WuYi province, with locals joining in a moving parade up the winding road to town, I was the American stereotype: naïve, but rather fun.

Positive Deviance (Fitting In)

If you are not fond of being categorized, it can be challenging to find a word that helps you fit into group identity. In our class on social change, we were introduced to the term "positive deviant." That is, "someone who has unusual solutions to complex problems, working outside the box of other people's restrictions." That sounds good enough for the artist in me!

Self-Identifying

In a social mixer at the beginning of a semester, the facilitator explained that he would call out a word and everyone who identifies with that word should go to one part of the room and stand together, leaving others in the group standing in their original place. Words included: woman, athlete, activist . . . This was energy-activating, but also complex and revealing. Language has so many interpretations. Sometimes what is labeled fun is anything but.

When choreographer Bebe Miller was rehearsing her work *Going to the Wall*, she used a similar technique to explore questions of race, identity, and gender. Addressing the context of the current, ongoing discussions on race and culture, the work featured a cast of eight dancers and used layers of text and movement. Within the ensemble, groupings assembled and dissolved in ever-changing combinations to illustrate the inherent conflicts and unexpected consensus which surround these issues. The piece began with Miller alone on stage, giving the audience time to closely observe.

As a *Dance Magazine* critic reported:

"So," she suddenly asks, "Now you know me, right?" You search yourself for an answer, your mind rattling off the obvious: black, female. Her dancers assemble like party-goers behind her, awkward, self-conscious. You're just getting over her initial query when a voice-over pelts you with more questions: "Who of you are looking for the gay men in the group? Who of you are counting the white people?" Then she dissolves into the gathering of dancers. A song from the hip hop group The Fugees bursts in and the party—or performance—begins.[3]

Bebe Miller and Angie Hauser
Bebe Miller Company
Photograph © 2008 Julieta Cervantes

safe, and requires less engagement. Yet, stereotyping may limit possible interactions by reinforcing our own collective identities and amplifying a protective fear of stepping outside our norm.

EMBODIMENT AND IDENTITY

Identities have deep emotional roots—people give up their lives to preserve them. Wars over religious and national boundaries and crimes from "othering" are ongoing. Volatility around collective identity permeates through the body, spans generations, and taps into inherited biases. It can perpetuate long-held grudges and distrust, often outlasting the original cause of conflict. Ethnic groups often share historical traditions and cultural norms through gatherings, ceremonies, customs, and dress and clearly differentiate themselves from others to preserve a strong collective identity. Being included or excluded from a group has implications. Peacemakers, experts in conflict resolution, and researchers focused on the psychology of ethnic and cultural conflict remind us of the powerful impact of collective identities on communication outcomes.[2]

Resistance may come up in relation to being categorized based on how others perceive us. Reducing life to fixed labels feels contrary to an inner, more expansive sense of self. Frameworks create power differentials. Although some people may want to insert you into a group context, fluidity between identities and stability in your embodied sense of self can offer increased range in communication and freedom from limited views and stereotypes. In conversation, for example, if you are a teacher speaking with someone who distrusts traditional education and is put off by talking to an academic, you can orient toward shared perspectives: you both have children, live in a crowded city, and care about water quality or equitable food access. Changing which identity leads in conversation can open doors; we are all human at the core. There is some place in which we can connect, even if it is simply standing on the same ground or sharing a personal story.

Passageway
Gemiler Island, near the city of
Fethiye, Turkey
Photograph by Nükhet Kardam (1995)

To communicate effectively with others across personal idiosyncrasies, professions, cultures, and stereotypes, we can notice the many-layered implications of embodied identity. Recognizing and appreciating our own multiple identities is a good place to start. As we notice which identity surfaces in particular situations, we can choose to utilize our whole-body intelligence to find places of connection. Inner work is involved. Rather than limiting ourselves to automatic defense behaviors, we can make conscious choices to engage, staying available for dialogue while defining healthy, resilient boundaries. Authenticity emerges when we feel "safe enough" to open toward another person, place, or circumstance. For some, safety is not to be assumed; there is a multilayered history that limits ease.[4] It is worth cultivating fluidity in identity as we interact in a global, changing world that is beyond our individual understanding.

Connections

Getting beyond your perceived personality can be a pleasure. Indulge yourself today by opening identity gateways. Explore your relationship to the many identity roles you engage throughout your day. Try something

Identity Signatures

Turkish educator and author Nükhet Kardam shares these stories about her bicultural identity:

"Traveling on a boat across the internationally significant Bosphorus waterway one day, I struck up a conversation with a Turkish man. After a few exchanges, he asked: 'You don't live here in Turkey, do you? Do you live in Germany?' I was talking to a man I didn't know. He was trying to fit me into a box, and I didn't fit. In general, a Turkish woman would have been more timid, sitting with her knees together, and not talking to strangers. Even though I am Turkish, my behavior, my confident body language, exposed my foreign identity. He was confused, and I was resentful that he didn't think I was Turkish enough.

"Working as a professional on the evaluation of a Women's Human Rights Training Project in Turkey, each of the stakeholders perceived me differently. To the women who received the human rights training, I was an American professor: I deserved respect as an older, educated woman from an unfamiliar world, but I couldn't possibly understand them. To the nonprofit organization that implemented the training, I was an outside expert: I had to be appeased and tolerated because their donor requested an external evaluation. To the teachers who delivered the training, I was someone who must be convinced that the program was successful: they had to hide any deficiencies. To the government bureaucracy dominated by men running the project, I was a woman with questionable expertise on a questionable topic, to be viewed with suspicion: greater gender equality was a threat and represented outside interference. Sustaining my core identity while fluidly moving between these multiple identities, I could respond to outer perceptions without defensiveness. It took conscious work and a sense of humor."

new, wear a different outfit, pause before responding to an inquiry. Try changing where you sit in a room, talk to a new person, spend time alone, and really listen to the world around you—not just to human voices—to refresh perception. You can so easily limit your possibilities for living and communicating by getting stuck in other people's views or categories—or your own. This is a day to expand horizons, unlock doors, and expand relationships. Anything is possible. A good day is a day when you are aware and inspirable in the present moment, open to interconnectivity within yourself and to others through more fluid and permeable boundaries and identity markers. ◉

TO DO
IDENTITY SHIELD
10 minutes

Identity is both a protection and a vulnerability.

Draw an identity shield. Start with an outer shape of a protective shield and use symbols and colors to represent the aspects of your identity that you choose to present. What are collective, professional, or personal symbols that reflect your sense of self or that others associate with you and factor into your sense of identity? Pause at some point to notice how your body responds when you distance from certain identities. Who am I if I am not . . . ? Take time to see what hidden layers emerge. Reflect on your process.

With a partner or group: Show your identity shield, and discuss its implications. Do others see you as you experience yourself?

TO WRITE
BORDERS AND BOUNDARIES
15 minutes

Writing can be like a fountain; you tap into a source (a creative vein) and it just keeps flowing. Write a series of stories about places you have traveled, borders/boundaries you have crossed. And if you are someone who prefers local terrain, write about places within that locale. Include how your differing identities are amplified or restricted in certain contexts, exploring the diversity within yourself. Keep the energy flowing in your writing, letting surprises seep through the cracks that might be limited by inhibition. When you think you have run your fountain dry, keep going past that perceived boundary; notice what lies beyond.

Spirituality and Embodiment

Day 23

Spirituality is ever-present. It is how we live a life. We can focus on creating the conditions for sacredness, and yet it is all around and inside us—the quality of light in a room, a baby's first cry, each intake of breath offering inspiration. It is up, it is down, it is dark, it is light. It is form, it is formless. Becoming aware of the pervasiveness of spirituality illuminates a view where "deep in the heart of matter, right here, right now and always, everything is interconnected, and all is well, and all manner of things shall be well." These words by Catholic priest and cultural historian Thomas Berry also remind us that spirituality does not separate; it connects. Interconnectedness is "not pantheism but pan-en-theism"—the mystery pervading and interpenetrating every part of the universe, extending beyond space and time.[1]

As a meaning-seeking species, many humans engage spirituality at heart as a questioning, a quest, an inquiry. There is so much focus on de-mystifying, quantifying, and assessing in contemporary life. Embodied spirituality is our invisible yet palpable way of sustaining connection to meaning and wholeness—holiness. It offers support for interconnectivity between the multiple aspects of ourselves, acknowledging mystery and not-knowing as part of present moment awareness. The journey is daily and life-long, personal and amplified in community. Spiritual traditions give shape to the ephemeral, creating containment and boundaries and offering their own answers or guideposts.

The body is our forest, our jungle, the "outlandish" expanse in which we are invited to let go of everything we think, allow ourselves to be stripped down to our most irreducible person, and see what, if anything, remains." — Reginald Ray, from Touching Enlightenment

Pecos National Historic Park, New Mexico
Site of a fifteenth-century pueblo and an eighteenth-century Spanish mission
Photograph © 2000
Louis Roberts

Behaving

My childhood introduction to religion was at a Methodist church in our small Illinois farming community. Our Sunday-school teacher was a hell-and-brimstone man who terrified children. He described a good angel on our right shoulder and a devil on our left as always whispering. He encouraged imagining suffering for our sins: nails pounded through the flesh of our palms and feet like Jesus on the cross. I often lay awake shaking, fearing burning in hell for a mean word to my mother or for a personal sensual moment. The body was bad, heaven was our ultimate goal, vigilance our job. Questions arose in my child-mind. Jesus on the cross: How could a father let that happen to his son? Father, Son, and Holy Ghost: Where are we girls in that triad? Reciting the Lord's Prayer, do I believe every word?

Fortunately, life on the farm and dancing offered different views. Miracles and radiance were daily occurrences: baby kittens born in my mother's closet, the exuberance of running freely in the alfalfa fields, and my humbling weekly ballet lessons in the nearby city where each difficult class ended with a grand reverence, a bow to the universe or to anyone gracious enough to be watching—like our mothers. The good part was that I felt seen in these multiple realms of childhood—religion, nature, and dancing. My behavior mattered.

The human body is the medium through which we encounter spiritual dimensions. We are spirit: the life force moves within us as well as around us. Words are symbolic; movement offers direct experience. Rather than rejecting the body in our quest for spirituality, we can consider our corporeal self as a resource, appreciating what we know from the daily experiences of our lives. Throughout history ritualized, action-centered practices often began by clearing and emptying—touching water to the forehead to bless, kneeling in respect, bowing into the process of beginning. Physical security supports psychological security. Within an appropriate context, any movement can emerge spontaneously from the body or be consciously endowed with a spiritual quality. For example, raising a hand can be a blessing or a warning; standing firmly can reflect grounded presence or defiance. With awareness, we can recognize seemingly pedestrian movement and gesture as meditative vocabulary within our daily lives.

Sometimes we have to rehabilitate our relationship to spirituality. While we genuinely feel the need for community, ritual, and connection to the mysteries of life, we can become dissatisfied with the contradictions between actions and words in spiritual traditions. Within formal repetitive practices, we can become "over-bodied"—digging ourselves into a hole by enacting movements mindlessly or repeating the same actions consistently, reinforcing old patterns. We may need to slow down enough to track more subtle sensations. Rather than telling the body what to do or being guided from outside, we can remain present to the unfolding mystery in the moment—the personal spiritual journey waiting to be evoked.

LOOKING BACK

Understanding the distinction between spirituality and religion can be clarifying. Religion—although complex to define—encompasses the rites, rituals, texts, and practices associated with specific traditions; spirituality is an embodied state that animates religious practice but also exists outside specific forms.[2] Ideally, they are so complementary that they become interwoven. Religious traditions are multiple and multilayered. The religions generally listed in history of world religion surveys include Baha'i, Buddhism, Christianity, Confucianism, Hinduism, Islam, Jainism, Judaism, Shinto, Sikhism, Taoism, and Zoroastrianism—and within each there are multiple sects and subgroups. Animistic and indigenous traditions abound around the world as well. All include teachings (oral or written) that name spiritual forces or enlightened beings—such as Allah, the Buddha, Brahman, Jesus, and Yahweh as well as Bondye, Olofi, and many others as they are named here and by names in different languages. They also codify embodied practices such as prostrations, hands together in prayer, and dancing. Within ancient traditions and contemporary practices, the mysterious dialectic between form and formlessness underlies spiritual experience.

Reflecting historically, we see that people since times immemorial have created teachings or doctrines about the wholeness of the world. Holy traditions were passed on to young people as a preparation for their future life in tribal communities as well as in complex civilizations. Three qualities seem to be shared worldwide that guide individuals to live and work together: ethical and moral components, silence and inward reflection, and service to the larger community. Once rooted in particular landscapes, spiritual traditions convey knowledge of nature, culture, and the proclivities of the human mind that have been gathered through the millennia. Individual wholeness was linked to a larger integrated worldview. Once a group of people loses a root connection to their core values, worldviews, and traditions (through war, genocide, political upheaval, cultural change, technology, etc.), vitality is drained and fragmentation occurs. Subduing a culture or colonizing a people, therefore, often involves disrupting or destroying unifying spiritual and cultural symbols and practices. In all eras, sustaining an integrated worldview can require a ferocious tenacity—even at this moment in our fast-paced, technologically distracted world.

Many spiritual traditions, including shamanic and mystical practices, were once carefully passed down and monitored from teacher to student.[3] Because of the potency of spiritual practices, there is a need to prepare: to enter the process effectively, to engage fully, and to return safely, integrating and embodying findings in daily life and communicating them to others. Relationships involve committed, intimate, and long-term trust between teacher and student as part of a lineage. Now, with instantaneous access to multiple formal teachings on the web, we have more choices, increasing the potentials for both understanding and misunderstanding. The risk is that we flit from one system to another, getting only superficial benefits or becoming lost in the process. Sustained concentration, a hallmark of meditative practices, is often the first personal discipline to go. With such easy access, fusion of multiple forms of practice can create confusion. Yet, the accessibility of in-depth wisdom traditions is also beneficial. It offers the opportunity to choose which practice works effectively at a particular stage in one's life.

One question emerges: to communicate about the ephemeral, do we need to use the same words?[4] There are multiple ways of addressing or naming spiritual forces and divine beings. Throughout history, humans have created many titles and symbols to identify that which cannot be named. For example, thirty-three types of deities are mentioned in the Vedic scriptures of Hinduism; five Dhyani Buddhas are icons of Mahayana Buddhism. This diversity of manifestations between and within specific traditions is inherent. The first chapter of the *Tao Te Ching* by Lao Tzu, a foundational Chinese text in Taoism, begins: "The Tao that can be told is not the eternal Tao. The name that can be named is not the eternal name."[5] In fact, the mystical dimension dissolves boundaries, engaging the interconnectivity of all spiritual practice.

The body of Earth can also be acknowledged as sacred terrain. Spirituality is often synonymous with light and transcendence, reflected in expressions such as "to be enlightened," "to see the light," and "to receive insight." Many ancient and contemporary architectural sites are aligned to create *hierophanies*—manifestations of the sacred through intentional light effects based on the position of sun or moon at different times of year.

Synchronicities

Synchronicities abound. Yet meaningful coincidences are easy to miss when distracted by outer goals. I learn this on a research trip to Hawaii when searching for Kuma Hula teachers to experience traditional Hawaiian dance. I visit the university, the Bishop Museum, and the Hawaii Heritage Center for connections and attend performances and workshops at area Halau Hula dance schools. Nothing feels substantial; I find no single artist for transmission. On the last day of my two-week trip, I receive a message from a friend: The master teacher on the island is living right next door. I never met her.

Writing About Spirituality

Immersed in words, I plan to take a break with a "Prana Flow" Zoom class by yoga teacher Shiva Rea. Somehow the connection number lands me in someone else's workshop. A didactic voice uses the word spirituality three times in the first minute. She promises a journey to my deepest self and mystical insights, employing the "lingo" of a spiritual guide. I do not want to listen to this voice and hope my chapter does not ring with these platitudes. But I can't get out of the Zoom meeting. After enduring ten minutes, I finally mute the sound and turn off my computer. You never know where you'll end up on the internet, but enlightenment is not likely to be found.

Sacred

"Sacred" is one of those words, like "soul," that requires more of you. Sacred has always held a shimmery quality in my mind, radiant. Yet visiting sacred grottoes or dancing amid the stone circles and rows in Cornwall, England, it's all about weight, wetness, and ground. People have held ceremonies at these sites through the ages, honoring seasonal and personal shifts, communicating with the mysteries of life. Soul, in my youth, also felt ghost-like, intangible. Yet soul music is all about heart and guts: getting down to the nitty gritty of life. I hesitate using these words in communication, feeling their span of meanings. Yet, some days a sacred tree connects me to soul—that is all there is to it.

I am driving to meet General Sherman. It is our annual West Coast pilgrimage to the giant sequoia at Kings Canyon, and this tree is, by volume, "the largest known living single-stem tree on Earth"—a beauty. I have talked to trees since childhood, telling them my innermost secrets and asking advice. Usually the response is simple—just one word (not to be shared). Maybe it is the spirits of my Nordic forest ancestors speaking, or the Nisse or Tomtes that are spirit-protectors of farmers, or my imagination. No worries; insight is insight. We receive it when and where we can.

Yet, the experience of the sacred also invites the dark. It draws us down to the roots, the soil, and the foundations of experience. Caves, grottos, and pools in nature are places of mystery, reflection, and inspiration, as are mountaintops, sky, and clouds. If we feel separate from nature and natural cycles, this disconnect may cause us to experience fear and insecurity, resulting in the desire to control or destroy landscapes, ecosystems, and other peoples, including their differing religious or spiritual views. Instead, when we experience ourselves as participants, we can resonate with the animate landscape. In many cultural contexts, spirituality, religion, and the arts are central, essential tethers to a coherent worldview. They involve nonverbal dimensions and value the complementarity of lightness and darkness and the universality of human challenges and joys.

TRANSITIONS AND GRIEVING

Life transitions often invoke spiritual dimensions. They are turning points, or liminal states, between old and new phases of life. These gaps in stability involve disorientation, hopefully followed by reorientation. With intention, we can remain open to what might happen in these times rather than shutting down or swirling in stressful thoughts. Surrendering into change requires relinquishment, descending into receptivity. Rather than tightening against disruptions, spiritual practices and religions offer multiple practices of reflection, silence, awareness, and cultivation of intention. They offer avenues to stay present and aware and to self-manage stress responses, bypassing potential illness and distress. Leaving your home community (willingly or by force), starting or ending a job, and the sudden diagnosis of an illness are some of the interruptive transitions that many of us will experience in our lives. Slowly, slowly we transition, whether we want to or not. You never know when life will change your direction, requiring something new of you.

Direct encounters with mortality often crack open one's heart and mind. The process of losing a loved one is a momentous, direct, and personal invitation to both grieve their passing and celebrate their life. How do we meet that occurrence, connecting to spiritual dimensions while staying embodied? Grieving is so mysterious and powerful that sometimes it can be too painful to hold and witness without community. Family, friends, counselors, or support groups may offer firm ground. Each process of grieving has the potential to open elevated states of awareness and vulnerability as we remember the sacredness of life. Grieving opens the doorway to other encompassing dimensions of despair such as grieving for the planet itself, since environmental destruction is so linked with life. One might hope that these moments and enduring experiences of ephemerality can enhance our capacity to care at every level for life. Rather than departing, escaping, or exiting the body, we can remain grounded and aware of the deep processes involved in moving through loss and helping our loved ones say goodbye to every treasured aspect of life.

EMBODIED SPIRITUALITY AND MINDFULNESS

Within each individual, spirituality engages a process of befriending who we are, staying present in the body, and not getting lost in judgment about ourselves or others. Deepening connection to our interior life—our inner spirit, essence, or guide—is a goal and a daily practice. The pathway to the self and to the heart is unique to one's experience, including insight into an integrated self. To be present in the moment, we have to feel. As we value our embodied experience, we listen more deeply. In-depth practice may involve a separation from one's daily life; turning toward inner work in a retreat setting, often in silence, can be conducive to uninterrupted descent, integration, and recovery. Clear parameters and appropriate guidance support the process.

Many mindfulness practices are not explicitly religious; they are awareness-building practices shared among and outside of spiritual and religious contexts. Embodied mindfulness practices like meditation, yoga, and qigong are compelling wellness companions that focus on awareness and discernment without attachment to preconceived outcomes. One foundational goal is to quiet the internal chatter of the mind, to feel more present to what is actually happening in the moment. Awareness of breath, which is always available in the present time, offers an inroad to meditative practices—an anchor for an overactive brain. Beginner's mind is a concept in Zen Buddhism known as *shoshin*. Zen master Shunryu Suzuki is quoted as saying, "In the beginner's mind there are many possibilities, but in the expert's there are few."[6] Opening to this fresh state of mind, even when you are well-versed in a subject, allows new perspectives. Rather than transcending physicality to engage spiritual dimensions, embodied practices locate transcendence in the here and now with every breath and choice we make.

Different *physical postures* are part of mindfulness training. Often, practices begin in stillness while sitting, standing, or lying down. It can be both humbling and exhilarating to recognize that someone, somewhere else in the world, is embodying the same posture with similar intent toward presence. Within the stillness of a held posture, there is awareness that each body is both universal and unique. Quietude is a relative experience; there is much going on inside the container of the body. In life and in our practices, we experience this flow between stillness and motion that can be infused with awareness.

Hands inform awareness. Because the hands have such extensive neurological mapping in the sensory region of the brain, they infuse responses and states of mind. Hands are signatures; fingerprints are personal to you as identifiers. Detailed hand gestures called *mudras* (meaning "seal," "gesture," or "mark" in Sanskrit) are essential components of Hindu and Buddhist ceremonies and statuary as well as utilized in yoga, Indian dance, and meditative mindfulness practices. Opening the palms has significance; we try so hard to hold on, grasp, control. Moving the hands—turning them upward, downward, or making a fist has impact, as does touching the

Water Rituals

Making a film beside the Pacific Ocean for a Global Water Dances project, we interview international students about their relationship to water rituals. Tangut, from Ethiopia, tells of carrying water daily from the river with her mother as a child, bringing a child-size bucketful to her father. Desmond, from Nigeria, falls to his knees with hands cupped, honoring and receiving. Leila, from Egypt, moves through a cleansing ceremony: washing arms, face, torso with imaginary water as she does five times a day in her Muslim purification ritual before prayer. So many of us have stories about water and its impact on the activities of our lives. Utilitarian and sacred dimensions merge.

In Ecuador, a guide takes me to visit seven sacred pools, ritual sites precisely oriented for solstice ceremonies centuries before. In Bali, we bring an offering to Tirta Empul, a water temple famous for its holy spring water where Balinese Hindus go for ritual purification. My own daily ritual is to submerge in a bath before sleep, warming, cleansing, inviting dreams. Rituals are repetitions, timely and timeless.

Bowing

In South Korea, before presenting a keynote address at a conference, the students bow to me as we pass in the halls. This honoring of teachers is humbling; it inspires me to elevate my awareness to meet their expectations. It is such a different response than at my current school in California, where the teacher is increasingly treated as a commodity: "Give me what I want."

Anjali Mudra
New Delhi, India
Photograph by April Danyluk

Transitions

As we refine and edit this chapter together, oncology massage therapist and dancer Susan Prins reminds me that spirituality is a component of massage; it supports the transitions from wellness to illness, illness to recovery, or living to dying. There is both a loving of bodily inhabitation and a grieving of transition. Touch can be a mediator—a way of focusing and attending to the ephemeral.

Words

There are many words about the invisible dimensions that animate and invigorate our lives. Yet none seem to capture the essence of the mystery that abounds.
It is like talking about a dance or painting: you never quite get it right. Colleague Bonnie Bainbridge Cohen affirms: "How can you use the word God? What else do you say? The Goddess? What else is there in the end?"[7]

fingertips in specific configurations. Try different hand positions; you will likely feel their interconnectivity with emotions and invocations.

Chanting is a practice of embodiment and transcendence in many spiritual traditions and community gatherings. Becoming the sacred word is a practice, a way of merging with gods, spirits, other people, and your own essential self. Practices that feature the voice and sound heighten awareness of the powerful nature of each word—both its vibratory tonal quality and its meaning. In the process of memorizing and vocalizing, holy texts and poems find their way into embodied understanding. Through chanting, singing, audible prayer, and recitations, you submit to the words—you become the images. Some texts and poetic formulae such as mantras must be spoken precisely and often; repetition allows the conscious mind to surrender—there is no need for vigilance. Sacred conversations and prayer are part of the lexicon of spiritual practices: *samvada*, a Sanskrit word translated as "a conversation, a speaking together," emphasizes the link between daily life and spiritual intention.[8]

Writing and *symbols* offer additional routes to embodied spiritual practices. Christian illuminated manuscripts, like the lavish Gospel books from the Armenian scriptorium at Skevra, depict angels whispering words into the ears of the writers, offering divine insight.[9] In some traditions, transmission of the holy words requires exact transcription, although translation from one language to another varies possibilities and meanings. Most good writing involves both the hard work of the author and the moments of inspiration that seem to come from other realms. In original contemporary writing, you often wait for the word to come through you, rather than from you: it is embodied and integrated rather than simply a mental, cortical musing. Some words, lines, songs, or passages appear unbidden and whole; you transcribe them immediately or risk losing their specificity and resonance.

Nature and *sacred architecture* evoke spiritual dimensions. Perhaps our deepest spiritual desire is to regard ourselves as part of the natural world, celebrating our participation in awesome energies and grandeur. The Japanese Shinto religion is a belief system in which every life form, object, and place is imbued with spirit. Animism in many cultures involves reverence for all living things—including non-living objects like well-used tools and shrines. The Lakota phrase *Mitakuye Oyasin* ("all my relations" or "we are all related") expresses this pantheistic worldview of interconnectedness and oneness with nature. Architectural structures such as churches, temples, mosques, shrines, and sacred kiva both contain and direct energies toward the divine, encouraging mind-body-spirit unification in communal settings while conveying specific meanings. Sculptures and paintings, textiles and carvings ground the ephemeral in lasting form and symbolic imagery. Re-enactments, pageants, rituals, ceremonies, and gatherings in designated locations help us access transformational states, offering a sense of cohesion and communion within ourselves and with those who came before: our ancestors who appreciated, imagined, or created beauty to celebrate and elevate spiritual dimensions.

Ecstatic states link trance-like movement to oneness with eternal forces. Deeply rooted movement practices can offer ways of inhabiting the body to stimulate mental states beyond the norm by connecting to invisible forces. Sufi whirling, Maasai jumping, Balinese trance dancing, Native American animal dances, and Buddhist prostrations are a few examples of ritualized practices that lead to transformational states of mind. Fasting and Christian traditions of flagellation and bodily exhaustion activate energies beyond those accessed daily, and contemporary extreme sports put life and death on edge, invoking heightened states of concentration. In our ordinary and extraordinary experiences of living and communicating, spirituality is "complementary to enchantment."[10]

Facing the Gods

Paul Matteson is performing a new dance called "how many times: bible backwards." As a religion and dance major in college, his reflections on dance and spirituality were always present. (He settled eventually on dance as his chosen profession.) Now, he tackles the words from the book of Genesis, speaking them backwards, one movement for each word, starting at the moment of expulsion from the Garden of Eden and backtracking to the beginning void. He scrambles things up; counterintuitive movement and words collide and intertwine. On stage, spirituality and corporeality battle it out before my eyes. At Paul's virtuosic level, he can embody almost anything—and does.

He contrasts this twelve-minute opus with a second dance, "how many times: bedroom stories," drawn from a three-month Buddhist meditation retreat. He recites *The Buddha's Discourse on Loving Kindness* (in Pali and in English) that he had chanted every night on retreat. This is now accompanied by impatient, desperate movements; interspersed, in a calm and soothing voice, he recites a children's story from *Frog and Toad Together*. He had both chanted the Buddhist discourse and read the sweet story to his newborn son every night for a year to encourage sleep, often in his own sleep-deprived state.

These stories we tell each other help make sense of our lives and calm our bodyminds. Facing off with the gods, the whole range of human experience is evoked: physical and mental are one. Spirituality, like child-raising and art-making, has an edge—they all require our best selves.

Moving Beyond

At the Somatics Festival 2019, Janet Adler speaks about movement: "Originally that's what all dance was, as far as I can understand: healing and mysticism."[11]

Connections

The invitation today is to notice how you address spirituality in your life and work. Honoring the spaces and conditions that hold potential for embodied spirituality, you can take time to choose and refine one personal ritual focused on conscious spiritual connection. You might create an altar, identify an inspiring place in nature, or include a mindful pause before beginning a conversation. In the face of the relentless stripping away of mystery in much of contemporary life, spirituality can be one way you establish, practice, sustain, and connect to a sense of meaning and wholeness. Although you may get more independent and particular as you mature, some part of you remains constant: your relationship to your endowed connection to the unknown.

TO DO

BRINGING DOWN THE HEAVENS (QIGONG)
3 minutes

In this practice, we extend upward toward the sky and then return that expansive energy inward to replenish our central core.

Begin in a basic standing posture with your hands by your heart in prayer position for emotional centering:

- Relax your hands downward and connect to the earth to replenish.
- Swoop your arms out to the sides and up overhead, gazing up toward the sky.
- Bring your hands slowly down in front of your body, palms facing toward your face, heart, then belly.
- Relax your hands downward to begin again.
- Swoop the arms out and up, gathering energy from the universe.
- Bring the arms down in front of the body, clearing and calming.
- Repeat, consciously connecting your thinking mind (brain), your emotional mind (heart), and your physical mind (belly) for a sense of coherence throughout your body.
- Repeat one last time. Sense and feel the expansive energy of the universe. Bring that energy down and distribute it through all the parts of the body.
- To finish, bring your hands together, palms touching your belly with thumbs crossed. Make a few circles with your hands on your abdomen and gather the energy there.

TO WRITE

ECSTATIC STATES

30 minutes to 2 hours

Often, dawn and dusk are times of heightened awareness. Take yourself outside at one of these transitional hours to write. Describe an experience where you felt enraptured, elated, transported, or euphoric—a sense of mysterious dimensions. Let this lead into a reflection on your personal relationship to religion and ecstatic states, beginning with your childhood and leading to today. Notice if your writing voice is different at this specific time of day.

Meditation

Everything is melting. As I sit writing, winter is turning to spring in Maine. Birds are returning. I feel the hard places in myself softening. This yielding reminds me of entering meditation. There is a softening of intention.

Adrea Olsen in *Awakening Grace*
Photograph by Alan Kimara Dixon

Day 24

Rest and Restoration

Times of chaos and challenge can be the most spiritually powerful ... if we are brave enough to rest in their space of uncertainty. — Pema Chödrön, from Lion's Roar *magazine*

Rest involves yielding. Although we do not need to micromanage this largely unconscious integrative process, we do need to value it by creating time and appropriate conditions for the body to replenish through sleep and deep rest. So much of each day involves "doing," but integration is about "being." Rather than more, more, more, we need time to "make sense" of what we do—to sort what to remember of our many experiences from what to forget and to re-establish connections that might impact decisions in the future. Rest and restoration are intertwined with resilient health through their abilities to fuel our communicative processes, vitality, and potential for creative insight.

The felt sense of rest begins with gravity: yielding weight. Experiences with bonding following birth as well as during events later in life—both positive and negative—underlie our capacities to release and trust. When the integrative parasympathetic nervous system activates, somatic signals or markers tell us we are beginning to relax. The breath deepens, skeletal muscles twitch and release, vital organs begin their processing—often

Undivided/Divided (2013)
Shen Wei Dance Arts
Installation at Yerba Buena Center for the Arts, San Francisco, California
Photograph by Stephanie Berger, courtesy of the Yerba Buena Center

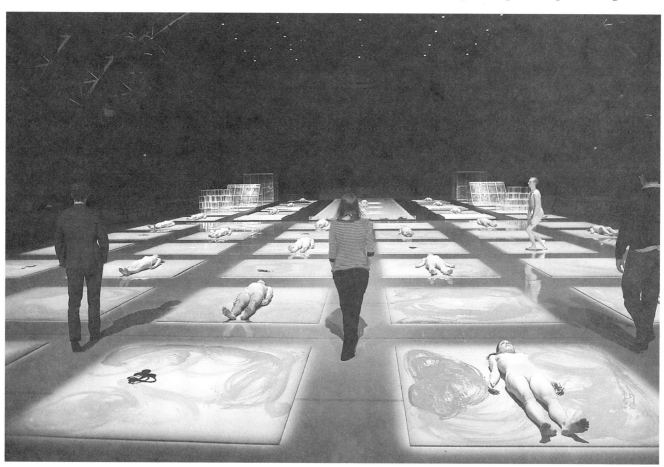

accompanied by stomach growls as digestive juices begin to flow. Yield when falling asleep is often described as dropping down, going under, or surrendering as you enter a less conscious state. Noticing these somatic markers—the sensations that indicate relaxing toward sleep—can help us replicate the process, stimulating the body's internal clock. Identifying conditions that support sleep even in novel or challenging situations, such as during travel, can assist in cueing restful states.

Sometimes we need to activate before resting. Especially if we have been sitting all day, working at a computer, or engaging in stressful interactions, it may be necessary for the muscles to release tension by stretching and moving to increase blood flow, clear stagnant energy in tissues, and flush stress hormones. Movement can be an essential precursor to effective rest. Meditation practices are often preceded by vigorous or flowing practices (like yoga or qigong) so the body is able to settle and focus. The relaxation response, articulated by Dr. Herbert Benson in his book by that title, encourages sleep by systematically contracting and relaxing muscle groups to return them to a neutral resting length.[1] In many cultures, a promenade or relaxing stroll follows dinner—benefiting digestion and helping to maintain healthy blood sugar levels rather than trying to sleep on a full stomach and overactive mind.

SLEEP SUPPORT

Because rest is so essential to clear thinking, it may be necessary to implement some "self-rules" to reserve sufficient time in each day and convince others that recharging serves everyone in the long run. We can value not doing, the empty space of allowing the body time to replenish. Rather than this personal rest being viewed as "selfish" or unsocial, in fact, it is generous to others to bring one's best self to events, conversations, the workplace, and interactions with family and colleagues. Recognizing that sleep is as essential as food and water, occupying about one-third of one's lifetime, we can maintain an effective daily plan to nourish, rejuvenate, and appreciate resting.

Good rest allows for good wakefulness. Going to sleep when we are too tired, full of the tensions of the day, is inefficient. The list of activities and repairs that must occur to replenish the body and brain while we are sleeping is significant; it is a wonder that so much can happen when we are seemingly "doing nothing." Tissues and body systems are serviced and toxins removed; metabolic processes impacting immune function, mood, and disease resistance are recalibrated and balanced.[2] Dreams synthesize experience and inform our days. Naps, meditation, and movement practices can cultivate a restful mind in the midst of stressful activities. Self-managing regular patterns of sleep and restoration supports the dynamic processes that affect how we function, forestalls a build-up of toxins, and prevents emotional exhaustion.

There are various dynamic stages to relaxing the brain in sleep. Our body's internal clock—the circadian rhythm—monitors both the intensity and length of rest in relation to environmental cues such as light,

Refreshment

Drained from teaching all semester at the college—giving, giving, giving—I attend a restorative yoga retreat at Kripalu Center. The first evening we gather pillows, blankets, props, and lie down on our mats. The lights slowly dim as we settle. A deep nurturing rest emerges through supported poses led by a skilled teacher in quiet community. Three days later, I leave renewed. This is a new baseline for my body, the "somatic marker" of sensations signaling deep release.

Flying to teach in Europe after an exhausting year, a row of seats in the center section of the plane is empty: I can sleep! So high up in the air, profound rest comes: dreams, twitches, imaginings, recovery. I travel from worry to refreshment in one transatlantic flight—just in time to land.

Collapse is another matter altogether. Giving up on a deep love, I lie on the beach near my parents' retirement community, belly down for hours, draining tension and grief into the sand. Late in the afternoon a man comes to me on the now-empty expanse, taps on my shoulder, and inquires: "Are you okay? You didn't move for so long, we wanted to make sure you weren't dead."

Water

Sleeping in the depths of the Grand Canyon on day four of a river trip, my husband and I are restless in the heat. We sit up on our open tarps under brilliant stars. Canyon walls on both sides amplify the powerful rush of the river and the night sounds of roaming animals. We talk about our life, dwell on problems. Edgy, thoughts begin to jumble and conflict. Every topic looms impossible: maybe we aren't meant to be together; maybe we should divorce . . . Then we remember our river guide's warning that if we start having strange thoughts, we are likely dehydrated. We creep down through the campsite, locate the picnic table with the water cooler, and drink our fill. Everything is fine. Body chemistry is real; when in doubt, drink water.

temperature, and perceived safety.[3] Numerous brain structures are involved: the hypothalamus (oversees the circadian rhythm), brainstem (transitions between wakefulness and sleep), thalamus (relays messages from the senses to the cerebral cortex), pineal gland (responds to darkness and produces the sleep-stimulating hormone melatonin), basal forebrain (promotes both sleep and wakefulness), midbrain (hosts attentional reflexes), and the amygdala (consolidates memory) are all active players. A complex array of interactions and processes across the body systems need ample time to process while we sleep.

Cues for good rest require preparation. Relaxing and quieting the mind releases alertness and vigilance, signaling that it is time to rest. By diminishing stimuli, we activate sleep cycles. A calming diet several hours before bed (less sugar and caffeine), plus reduced screen time and dimmed light sources (less eye stimulation), help to prompt easeful sleep. Quiet and darkness support natural circadian rhythms. A familiar schedule; bedtime rituals; massaging the hands, feet, or face; and considering a list of appreciations for the day to move the mind toward a restful state rather than dwelling on problems encourage good rest. If you have trouble falling asleep, remembering other situations that promote deep breathing—like visualizing a favorite, safe sleep location—can let the body relax.

DEEP REST

There is a difference between yield and collapse. *Yield* is a release of weight into the ground; we eventually feel a natural rebound of energy that allows us to move away with ease when we are rested and refreshed. *Collapse* is a draining down of fatigue toward the support of the earth that can feel endless if deep and prolonged exhaustion is present. Sometimes it takes a long time to feel the spontaneous impulse to rise; there is so much to let go of, so little energy available, so much to absorb. The earth is a natural recycler. Yielding weight and allowing oneself to collapse, when necessary and in a safe context, can deeply rejuvenate and revitalize the body.

As we release sympathetic nervous system activation and move into parasympathetic integration and digestion, we promote ease and equanimity in the whole body. Deep rest can impact many conditions, including mild depression, high blood pressure, headaches, ongoing fatigue, and insomnia. Resting during major transitions allows the body to refuel and reframes priorities; these are necessary so we avoid illness. Even after a celebration, success, or spontaneous insight, there is a let-down cycle that requires time for supported rebound.

Lethargy is the shadow side of rest—you just can't get going. Beyond depletion of energy, lethargy and sluggishness can reflect grieving or hormonal imbalances that perpetuate ongoing fatigue and lack of enthusiasm and inertia. On the other hand, overly activated energy can result in restlessness that also signals distress. In both of these polarities, certain body systems are compromised or overstimulated and in need of attention and regulation. Retuning to natural cycles can help—sunlight, water, and seasonal shifts offer support. Walking, swimming, bicycling, paddling, gardening or any invigorating connection with nature lets us feel part of rather than separate from the larger, elemental world.

Outer structures in our lives can establish space and security for rest. For some, a formal religion, marriage, good job, nature, home, family, or a sense of purpose and usefulness allows parts of the overstimulated mind to rest. More inwardly initiated supports for rest include a daily movement or meditation practice, music, nourishing meals, and self-initiated rituals. There are so many ways to find ground in the flux of our lives. In some periods we need time alone; at other times, resting in community or in the presence of a teacher or mentor offers more support. Both silence and sound are valued—containment and deep listening as well as expression and engagement. Two questions emerge: "How much structure do I need in order to rest?" and "What are the conditions that support my deepest ease?"

INNER VOICES

There can be an inner voice that interrupts, saying "WORK HARDER." Psychologist Carl Jung described this resistance to rest, especially in women, as a relentless *animus* drive—an inner masculine part of the female personality. It commands us to do more, be more, and strive more aggressively to achieve or to support others.[4] The animus in this context does not care about one's individual health; bearing exhaustion to achieve outer goals (feeling busy, useful, successful) is a component of a strong animus drive. Although many parts of the brain and body systems have "brakes" to monitor too-muchness and maintain homeostasis or balance for survival, the psychological drive within humans can be relentless. Often it is the body itself that breaks down with illness or injury as a warning to stop, rest, and recover. Giving oneself permission to relax is fundamental. It is more effective and life-enhancing to "tune up" and "tune in" before illness sets in. Clearing physical, emotional, and mental blocks, clarifying intentions, and valuing rest as a cyclical component of a fulfilled life encourages integration.

Time has impact; varying conditions require healing sabbaticals of different lengths. Setting aside an hour each day for an integrating class or personal practice, a day each week for rest and relaxation, or a month-long retreat for recovery, with appropriate support, can reset one's inner clock and signal new beginnings. Valuing time to recover from the work of the world can reset healthy patterns. Prayer, the Sabbath, sabbaticals, vacations, and mindfulness retreats are a few of the ways we designate time for reflection and integration. With virtual media literally at hand almost everywhere we go, sometimes we have to "unplug," take news breaks, go "off the grid" to experience real ease. The relentless drive for constant stimuli is part of addictive behavior. If one cannot let go of something, it is a warning. Whether this is morning coffee, repetitive thought patterns, relationships, or checking the cell phone, taking a break means there is freedom from unconscious drives. Healing dependence on drugs, alcohol, medications, and other deeply embedded neurochemical circuits, on the

Safe Company

Hosted by a Māori family in New Zealand, our dance company is taken on a special visit to a *marae*—a Māori "gathering place" and carved meeting house. As we approach, our host stops the van so we can pick fresh watercress in a clear stream for a present. On arrival, we are greeted by the community and exchange welcome dances on the open ground outside before entering. Removing our shoes, we pass into a spacious chamber; it's like walking inside a human body. Carved panels tell the *whakapapa* (genealogy) of the tribe, detailing Māori stories and legends. We dine, then sleep together on mats spread around the dark wood floor—no walls. Our host explains: if we dance together and sleep beside our guests, we become friends, not enemies.

Resting Places

There is a place in Wales where I go in my mind when I want to rest. It is down a steep hill, along a path, by an old-growth forest. Soft mosses—unimaginably green—cushion each step. I lie down next to a pond amidst tall waving grass that masks me from view. It is warm but not hot, and the field of clover nearby houses the faint hum of bees. Visualizing this place where I have relaxed so deeply in the past creates all the cues in my body for ease. I begin breathing deeply, as though I am still held by that place.

other hand, can require trained support from professionals to return to balanced life.

RESTING FROM TRAUMA

Definitions of and responses to trauma have broadened with popular usage of the term. We can consider trauma as an interruption of flow or a fragmentation or breaking off of self from an integrated whole. Healing involves re-establishing credibility with your psyche: trusting yourself, which supports the ability to trust others and the physical world in which you are embedded. Resources and supports are essential within the process of restoration toward a unified self—not just a return to how you were but as you are now in this new present moment. *Titration*—the process of avoiding overwhelm of too much, too fast—can be essential when working with deep-seated trauma patterns.[5]

Movement practices can be helpful to discharge excess tension and build resilience. Time and timing are factors; speed and tempo play important roles. Faster movements help with dissipating disruptive energies, cleansing and purging the body systems. Slower, integrative movements help to calm and cultivate restorative energy, returning healthy tone to the system (not hyper or hypo-tonic). Moving energy in the body into a state of optimization—healthy, enlivened, resourced—we can meet challenges through stability and a safe-enough container. Generally in our daily movement practices, we need a range of expressive and restorative periods that support internal boundaries. *Pendulation*, or arcing, between these two polarities involves developing the capacity to both actively engage and restfully digest. Ending movement practices with a calming, integrative state—building energy reserves and establishing resources for our next encounters—is part of valuing rest, restoration, and integration in our daily life cycles.

Each day, each moment, and with each breath, we have the opportunity to rest and restore. We can value unscheduled, unprogrammed time and also build in frameworks for relaxation and healthy sleep patterns. Our attitudes and values around rest impact our inner stability and security and broaden our opportunities for integration and exchange. Much is happening that is essential to the health of the brain and body while we sleep; deprivation creates serious depletion. Effective living and communicating means bringing our best selves to the process. Sometimes this involves taking time to know ourself more deeply and being a good friend to the parts of ourself that need rest, integration, and restoration for a sense of embodied wholeness and well-being.

Connections

In this day focused on rest, you can notice when and where there is time and space to rest and refresh. Invite a few breaks in your usual routine to change gears and recalculate tension levels. What feels restful to you? You are the guide, the driver of your day. Identify one simple, restful practice to self-manage tension. This might include shaking out excess or stagnant energy or a qigong practice like "Bringing Down the Heavens" (see Day 23). Sometimes rest requires yielding; if you are already feeling lethargic or exhausted, you may not want to give in. But valuing the cycles between rest and activity several times during your day can help you be more productive and present, revitalizing each moment. Notice how recognizing the interconnectivity of rest and restoration during your day can balance and refine an ever-active, over-active life.

TO DO

CONSTRUCTIVE REST
10 minutes

Constructive rest is an efficient position for re-establishing neutral body tone, allowing the skeleton and organs to release into gravity and encouraging deep relaxation.

Lying on your back on the floor, in a warm, private place:

- Bend your knees, and let your feet rest on the floor, hip-width apart.
- Let your knees drop toward each other to release your thigh muscles.
- Rest your arms comfortably on the floor below shoulder-height or across your chest.
- Relax into gravity; allow yourself to be supported by the floor.
- Feel your breath and the responsiveness of the whole body.
- Allow the organs to rest inside the skeleton (the lungs and heart, digestive and reproductive organs); feel the contents release within the container of your skin.
- Allow your brain to rest in the skull.
- Allow your eyes to float in their sockets.
- Allow your shoulders to melt toward the earth.
- Allow the weight of your legs to drain into the hip sockets and feet.
- Allow the surface of your back to move against the floor as you breathe; feel your ribs articulating at every breath.
- Allow your jaw to gently fall open; feel the air move in and out through your lips and nose.

As you release your body weight, the discs are less compressed, and the spine begins to elongate. You may need to lift your head or pelvis and lengthen the spine on the floor to accommodate this change. There is a natural condensing and expanding of tissues, allowing flow throughout your structure. Constructive rest is useful at any time of day, but especially if done for five minutes before bed so you don't sleep with the tensions of the day.[6]

Longing
Bonnie Bainbridge Cohen offers: "Rest is reducing intention."

Fabiano Culora
Marloes Sands Beach, Wales (2015)
Body and Earth: Seven Web-Based
Somatic Excursions
Still image from videography
by Scotty Hardwig

TO WRITE

RESTORATION (YIELD OR COLLAPSE)
20 minutes

> *Sometimes restoration comes by yielding and resting until there is*
> *a natural rebound to stand and move on. Other times, if you are*
> *totally drained, you may collapse to refuel, finding essential time*
> *and support to regain equilibrium.* Describe an experience of deep
> exhaustion; what did it feel like to need to yield into collapse and
> regroup? What conditions made it possible? How might you invite
> more restful states into your current life to keep exhaustion at bay?

Food and Ritual

Nourishment and nurturance are interwoven. We are born with a heartfelt relationship to food. In the womb, nourishment arrives through the belly, but once we are "airborne" the mouth is the entranceway. Whether breast or bottle-fed, we are cradled and comforted on the chest of our caregiver near that familiar rhythmic pulse of the heart. How that early relationship to nurturance linking food and love is cultivated or disrupted underlies one's personal food story. Rituals and celebrations around food are extensions of one's broader community, including its multicultural dimensions. Everyone embodies a heritage around food that is worth investigating in a desire for more easeful communication with ourselves, other people, and the places we live and love.

Delicious food shapes us—it is the very structure of our being. When sufficient food is available, finding the right nutritional balance fuels our capacity for interaction and enjoyment. Imbalance, on the other hand, fosters distraction and disconnection. Food attitudes are impacted by familial, social, and cultural imprints; they also reflect an individual's own body chemistry including allergies and food sensitivities. We each have the capacity to create the conditions in ourselves for peak engagement. Food, as a chemical elixir sets the tone, literally and figuratively, of the body.

It seems to me that our three basic needs, for food and security and love, are so mixed and mingled and entwined that we cannot straightly think of one without the others.
— *M. F. K. Fisher, from* The Art of Eating

Labels

We walk into our busy supermarket in northern Maine and see that one half of one aisle is marked "Nutritional Foods." What, we wonder, do all the other aisles contain?

A Swan's Wedding Day (2008)
Sculptural teapot by Mara Superior
High-fired porcelain, ceramic oxides, underglaze, glaze, wood, gold leaf, bone, ink, brass pins, 21 in. × 16 in. × 10 in. Renwick Gallery, Smithsonian American Museum of Art, Gift from the Kohler Foundation, 2019.7.1

Food Connectivity

Local organic produce deserves special attention! It feels good to be part of the interconnectivity of people and place. Tonight I'm making rosemary chicken, with fresh herbs delivered by a neighboring cook and chicken from Carly and Aaron at Tide Mill Farm (who care for land that has been "home and workplace to the same Maine family for two hundred years—nine generations.") I look up the recipe in my hand-written cookbook, tattered and falling apart from four decades of use. Meaningful quotations are interspersed with party descriptions and favorite recipes. For this one, I slather the sauce (involving Maine honey, local mustard, last year's garlic, and—I confess—ketchup) over two fresh half-chickens, to be roasted alongside homemade French fries. Rosemary chicken was a standard meal in grad school; my cookbook entry from 1972 says: "Messy but flavorful!" The juxtaposed quote, well-decorated with food stains, is from writer Saul Bellow: "The great virtue in art is that it delays you." Below the recipe is a reminder to listen to the piano sonatas of Poulenc. Food, conversation, and ambiance, all recorded.

Eating can be considered a ritual; we repeat food habits several times a day, wiring them into our systems. If we define ritual as "a repetition, an act or series of acts regularly repeated in a set, precise manner," we recognize that both consistency and intention are involved in preparing and sharing food.[1] How we relate to ourselves, the environment, and the food itself is part of the process. The pleasure of food and the ceremony of dining—of meals shared or alone—has implications in embodiment: we are "what" we eat. Underlying assumptions have impact: we can consider a meal as a source of conversation and communion or as something to get through as quickly as possible—eat on the run, or pause and enjoy. Choice is required.

Food is a cultural signature. Social interactions with food are layered with ancestral practices and seasonal celebrations. Economic implications are a factor in contemporary life choices—what traditions did we inherit, which foods can we now find and afford? In international settings, we can engage local cuisines: savoring the scents, flavors, diverse food combinations—spicy curries or unusual fish—and ways of preparing. By participating in different social patterns of dining, we can reflect on the relationship between connectivity and longevity. Knowing that many in the world and in our home communities are undernourished and have limited healthful choices, we can listen carefully to the food stories around us with respect, curiosity, and informed concern.

FACTS

Digestion is a process of incorporating the nutrients of the earth into the tissues of our bodies. Occurring successively in each digestive organ that forms our body-core, digestion is a process that begins with the nose, eyes, and tongue. The sensorial choice of whether to accept or refuse food involves the smells, colors, temperature, and textures presented. We can watch babies pursing their lips to keep certain foods out, opening their mouths enthusiastically to receive, or frowning and spitting out unfamiliar textures or tastes. Once accepted by the mouth and chemical-sensitive tongue receptors (taste buds), the food is swallowed and continues on its journey through the *esophagus* linking mouth and stomach. Because this long tube passes between the lungs, behind the heart, and through the breathing diaphragm, "heartburn" is a common term for indigestion occurring in this sensitive region.

The primary function of the expandable *stomach* is to break down food using strong digestive juices. Many people don't really understand the location of the stomach, which is on the left side of the torso beneath the diaphragm and ribs, not in the belly. (The stomach can extend all the way down to the pelvic base after a big meal, however, moving other organs out of the way.) Next, the food passes from the stomach through approximately twenty-one feet of *small intestine*, where 90 percent of nutrient absorption takes place over a period of three to ten hours. The *hepatic portal vein* carries blood (containing nutrients, medications, and toxic substances) from the gastrointestinal tract, gallbladder, pancreas, and spleen to the liver for blood filtration before cycling the nutrient-rich blood to the heart to be

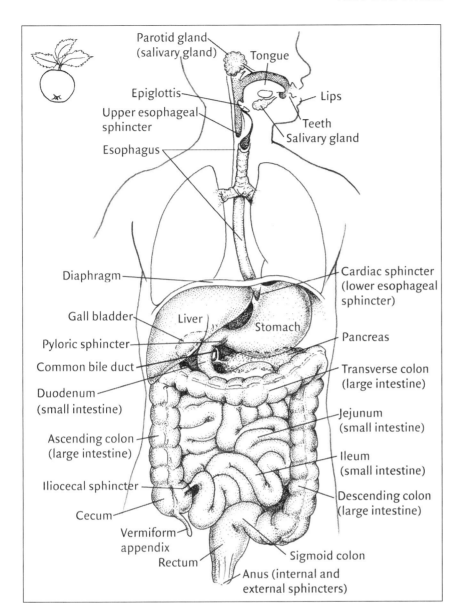

Parotid gland
(salivary gland)
Tongue
Epiglottis
Lips
Upper esophageal
sphincter
Teeth
Salivary gland
Esophagus

Diaphragm
Cardiac sphincter
(lower esophageal
sphincter)

Gall bladder
Liver
Stomach
Pyloric sphincter
Pancreas
Common bile duct
Transverse colon
(large intestine)
Duodenum
(small intestine)
Jejunum
(small intestine)
Ascending colon
(large intestine)
Ileum
(small intestine)
Iliocecal sphincter
Descending colon
(large intestine)
Cecum
Vermiform
appendix
Sigmoid colon
Rectum
Anus (internal and
external sphincters)

Digestive tract, anterior view
Anatomical illustration
by Nancy Haver (2021)

circulated to all the body's cells. The *large intestine* is mainly concerned with absorption of water, minerals, and some vitamins, moving the remaining non-nutrient residue toward the final *anal canal.*

The digestive process can be efficient or compromised. Along the way are many important junctures, which can be places of holding or easeful transition. Seven muscular rings, or *sphincters*, link one part of the digestive tract to the next, beginning with the lips and followed by the upper esophageal (linking throat to esophagus), lower esophageal (esophagus to stomach), pyloric (stomach to small intestine), ileocecal (small intestine to large intestine), and finally two anal sphincters (rectum to outside the body). There is a choice at each juncture whether to go forward or to hold on, reflecting larger implications of taking in, integrating, and letting go. Our attitudes about food affect all aspects of the experience. When the nourishment cycle is completed, digestive processes return unused nutrients back to the earth.

Dining Training
While living for three months in Whitby, New Zealand, we are "tutored" by a small local restaurant about dining customs and expectations. When we reserve a table for two, the host tells us they have one sitting per night: "No matter what time you arrive, the table is yours for the evening." That should have tipped us off. Our waiter looks us over with scrutiny. We are slow to place our order, and then we wait—chat about our day, look at our guidebooks, and wonder how to fill the time. When our meal arrives, each part is stacked: riding atop the New Zealand lamb is a swirl of mashed potatoes, a bit of beet salad, and a towering decoration of onion rings; a drizzle of balsamic reduction decorates the plate. Delicious. When we finish "tucking in," our waiter informs us we are to place both fork and knife on the plate to show we are done—even though there's not a scrap of food remaining. When my husband asks if there is soy milk for his after-dinner coffee, he is assured, "No problem." Then we see our waiter driving away on his motor scooter to buy some. The evening grows longer! The take-away message: talk to each other; do not be in a hurry—this is dining, not eating.

Bests

My best seafood lunch ever is in Bari, Italy at the restaurant *Al Pescatore*. Our outdoor meal is selected by Italian dance artist Elisa Barucchieri; what she chooses is fresh— some raw, some cooked—with multiple courses. We start with the raw seafood *antipasto*: sea urchins, oysters, mussels, calamari, and shrimp straight from the sea. Then we consume a mixed seafood grill for *secondo piatto*. When a man from a nearby table sends over a bottle of wine, we move into what my colleague at the table calls "recreational eating"—you have been eating for pleasure and are now full, going into overtime enjoyment—too full! Dining in Italy is an event: you'd better show up hungry.

In Avezzano, located in the mountains of central Italy, we feast on home-grown produce with Susanna Recchia's family. Tasting sweet black cherries right off the branch and dining on fresh chicken delivered by a neighbor, *burrata* (made from mozzarella and cream) cooked on an outdoor fire, truffles found with their well-trained dog, and aged balsamic vinegar on our fresh greens, we celebrate garden, orchard, forest, and field. In this region they maintain the ten-kilometers rule: no olive oil, wine, or produce from beyond that distance comes to this table. We leave with small jars of golden saffron, the crimson stigmata and styles from crocus flowers cultivated and gathered by her father. And it is not just the food; the conversation and company are delicious too.

By understanding the digestive tract, we can visualize food passing through our bodies. At each digestive organ and muscular sphincter, we can encourage relaxation and movement through the system. Basically, the mouth and stomach break down food, the small intestines are the site of nutrient absorption, the accessory organ of the liver filters the blood, and the large intestine reabsorbs fluids and prepares for elimination. Touch and hands-on massage can be used, especially around the sphincters, to move food through the tract. Deep breathing or a relaxing afternoon walk can activate sluggish tissues and fluids, assisting the dynamic process of nutrient flow in the body. Fluid intake, frequent urination, and regular defecation are essential to remove toxins, although too much fluid can also flush nutrients from the body.

In the lengthy digestive process, we can consider quiet time as the most vital, integrative aspect of our days and nights. Many cultures build time for digestion into the rhythm of the day with siestas or restful periods following meals. Dining is also a natural gathering time for families, allowing conversation and exchange about the events of the day. When we plan meetings at lunch and allow our family rituals to be disrupted by phone calls, electronic devices, sports practices, and long work hours, we are sacrificing essential connections such as relaxing, focused listening, and digesting. Try building at least one leisurely meal into your week, and notice its implications for your relationships and your health.

Digestion is closely related to other integrative functions in the body, such as emotions. When we are in the midst of a difficult or new situation, the digestive tract may reflect our uncertainty. For example, if you are arriving in a new culture, you may have difficulty taking in the situation and feel a "lump" in your throat or lack of interest in eating. Sometimes you may be able to take in a new experience or difficult encounter but resist integrating it into your life, resulting in indigestion—queasiness—or stomach ulcers. Sometimes you can integrate and digest an event into your life but resist letting go of what is no longer essential from the past, resulting in constipation, diarrhea, or colitis. New foods and cultural cuisines may also impact digestive efficiency and comfort levels. Not all digestive problems have emotional correlations, but it is worth considering emotional implications if the digestive tract is in turmoil.

In balanced alignment, the skeletal system supports digestive function. Bones transfer the body's weight to the earth so that the digestive system and all the vital organs are free to function without compression and restriction. With effective skeletal support, expansion and condensing of all cells and tissues occurs, along with wavelike constriction and relaxation (peristalsis) in the digestive tract to transition the food through your system. Holding or over-contracting the abdominal muscles (abs of steel!) restricts healthy activity; on the other hand, making space by effectively aligning the spine without shortening the front surface (slouching or actively flexing) or arching (hyperextending) enhances energy reserves. Dynamic balance in the body involves finding skeletal support and equalizing muscle tone through whole-body awareness for easeful organ function.

COOKING AND DINING

The experience of eating can inspire appreciation. As Michael Pollan writes in his book *In Defense of Food: An Eater's Manifesto*, "The shared meal elevates eating from a mechanical process of fueling the body to a ritual of family and community, from the mere animal biology to an act of culture."[2] For creative cooks, preparing food—choosing raw ingredients, deciding what to make of them, and cooking alone or together—is part of the delight and ritual of dining. Many people find cooking as satisfying as eating. Collaborating on a meal both affirms relationships and enhances connection to one's food. Relationships with where our food comes from integrate and implicate us in regional and global food systems. Supported by local grocery stores, outdoor markets, and personal gardens, preparing food links us to other households and communities in the daily process of creating meals.

Cooking is a kind of alchemy. As in all creative processes, there are steps and stages. Assembling a meal or a special dish includes selecting favorite recipes and ingredients, shopping and assembling, combining and cooking, tasting and timing, and plating and serving, along with other preparations like arranging the table. All of the senses are activated: the aromas and visual delight of what the meal looks and smells like, the sounds of conversation, and the combination and juxtaposition of courses around which friends and family gather. Great cooks have an intuitive connection to the entire process and a passion for creating and sharing that is about not just the food itself but the context in which it is offered. Often, there is collaboration and support along the way—willing family members or friends who absorb cooking and eating skills as well as joining in the pleasure—linking us to the moment and creating memories for younger generations.

Eating can be an adventure; food is infused with memory and associations. We can travel the world in pursuit of new tastes, customs, and visual delights or enjoy distinctive meals or flavors right in our own town. In Hawaii, for example, where a diverse array of cultures coexist, one plate of food may contain tastes from multiple heritages, countries, and cultures. In some contexts, sharing food is a great honor, building trust and relatedness. In Turkey, tea or coffee is served before any serious conversation, creating the basis for more in-depth connection. In China, the shared meal is placed in the center of the table and diners pick up their chopsticks and dive into steamed chicken with ginger sauce, tasty fried noodles with vegetables, hoisin sauce ribs—so many choices.[3] In Ethiopia, friends and family often tear off a strip of bread, *injera*, to wrap around foods like meat or curry and put this in the mouth of a friend or family member to show affection. There are many dimensions to consider globally; cultural connectivity is part of dining.

GROWING, GATHERING, AND SHOPPING

What *is* food? The juicy tomato or fresh peach dripping down your chin are from edible plants containing chlorophyll, which convert energy from

Visits

An Eritrean student in California describes a visit with her grandmother and their traditional coffee ceremony. This multi-stage Eritrean custom involves her grandmother going to the kitchen to freshly roast coffee beans and sampling the smoke to absorb the scent, followed by grinding, boiling, pouring, serving, and filling one extra cup (for a mystery presence). The custom is for a guest to refuse the coffee three times before accepting. This is just the beginning! The brewing is repeated over the coffee grinds four times. Contrasting her acquired Western time-sense with the current familial situation in her homeland, she has forgotten what a reunion with relatives involves—there are no short visits.

Choices

At a residential retreat near the Yorkshire Dales, our fantastic cook Julie Nutchey brings a stack of cookbooks to accommodate multiple food "needs." It is a weeklong gathering featuring vegetarian fare for a group of fifteen. Dietary restrictions require wheat-free, dairy-free, and sugar-free options. Serving three meals a day, she has a challenging menu to prepare and label. One evening, we are all seated outdoors at a long table for a celebratory evening dinner, and she serves gluten-free, non-dairy chocolate cookies with mounds of whipped cream topped with maple syrup. Yum! Everyone wants some!

A friend, a fine cook, invites us to dinner, stating: "I serve what I serve, and you eat what you eat." No accommodations: the responsibility is on the guest to pass up one of her delicious dishes—or to enjoy.

"Blanche, Ethel, Feona, Loretta, and
Mr. Bates" (2018), pillow by Cathy Shamel
Original design in hand-dyed wool

sunlight. This family of vegetables and fruits provides the base for our favorite treats like coffee, teas, and chocolates, plus seeds and nuts; tree-ripened fruits like apples, pears, and ruby-seeded pomegranates; the hearty multicolored vegetables that fuel our meals; and the spices and flavorings that transform basic food into delectable cuisine. Edible mushrooms, molds, and yeast are a category in themselves—fungi. And nutrient-dense seaweed is a category within the broader umbrella of algae. These are all part of the first "trophic" layer on the food web, primary producers that give humans the most direct access to energy from the sun.

Food includes animals—fish, birds, and mammals that eat plants or other animals. As nature writer John Elder says in a class on local foods: "Nature is something that we eat and eats us." Various meats, seafood, milks, cheeses, and other animal parts and products support many people's lives around the globe. This involves the second and third trophic layers: herbivores like grass-eating cows and deer are once removed from the sun's direct source of energy; and carnivores that eat other animals are in the third trophic level. Humans are generally omnivores, eating both plants and animals, and are slightly removed from the original energy source of the sun. Minerals play a role in our diets: salt, calcium, and other added elements give flavor, texture, or longevity to food (not necessarily to us!). Beyond these natural components, there are food substitutes, fillers, and additives like chemical compounds that lengthen shelf life. These may hinder the natural process of digestion, causing adverse effects to the pleasure of dining, but they also support those who live far from food sources.

Growing and nurturing our own food is rewarding; it is also hard work, requiring skill, knowledge of soil and weather patterns, and a healthy dose of humility when we cannot control all of nature's elements. It is a revelation to watch a seed, when planted and nourished, become a plant that responds to sunlight, water, and care—or witnessing an egg become a chick. Picking a warm tomato right off the vine, pulling a bright orange carrot from the soil, or gathering green beans to share with friends offers deep satisfaction. Many children living in urban centers are cut off from the life cycles of food. We teach now what we used to take for granted, including that energy in our bodies comes from sun and soil. Growing plants in our homes can be helpful; we develop a relationship with their needs and their contributions to our lives. As we clip fresh basil and savory rosemary from plants in our kitchen, we know that the whole household benefits from their pungent aroma, air purification, oxygen production, and flavorful presence in our food.

Shopping for food can be a pleasure, especially when relationships are established with local vendors. There is a social component to knowing the farmers who supply your markets, choosing which stores have the best selection, talking with the butcher for valuable knowledge about cuts and freshness, frequenting the cheese and wine store for local flavors reflecting the *terroir*—the taste of the land—or cultivating a relationship with the baker for special holiday treats. Where local markets have been displaced by fast-food and chain restaurants, creating a "food desert," shopping can be challenging. Supporting regional farmers and food networks encourages self-sufficiency in the communities we inhabit. Where we spend our time, money, and energy can be infused with goals for health, pleasure, and multigenerational, cross-cultural communication.

STRESS AND DIGESTION

Dining can be stressful. As we work too many hours trying to afford "healthy" food, while consuming nutritional supplements to boost or detoxify our systems, the simple pleasure of eating can be laden with complexity. We often encounter variations in cultural and regional customs around dietary needs and contradictory information about nutrient values. Differentiating the benefits, for example, between "local" and "organic" produce can be confusing; yet for many, finding enough food is a daily challenge. In the media, we are shown unrealistic views of body image—impossibly thin or overly strong—and hear troubling stories about the health effects of preservatives, pesticides, herbicides, hormones, genetically altered food, and the chemicals in processed foods and our water supply.

Ethics are implicated: Who is making money on this product? Who is growing the food and how are they treated? What is the relationship between food production and soil health in the region? Or we may be faced with a plethora of dining possibilities, restaurant menus offering multiple cultural cuisines with delicious options and unfamiliar names. Or eating in a college cafeteria or on a luxury cruise with way too many plentiful offerings. We can be consumed with anxiety and the challenge of too-muchness or guilt around consumption.

Digestion is interrupted during stressful encounters. The body puts all its resources toward activation and defense. We waste energy and nutrient intake that could be used to rebuild and refuel the body's systems. When threat is perceived to be present, the *sympathetic* portion of the autonomic nervous system activates vital resources toward survival. (Heart rate increases, skeletal muscles flush with blood, pupils dilate—it's exciting!) Secondary processes like digestion and reproduction are put on hold until danger is over. Planning and complex thought-processing are also secondary. Unfortunately, we in contemporary industrialized cultures are often reacting to every external stimulation as a life-threatening event, wasting precious energy resources in ways that do not support our well-being. Skills to cue parasympathetic relaxation and integration support creativity and insightful reflection (see Day 9).

Yet it seems that many individuals are habitually distracted (busy!) during meals: if we do not pay attention, we can pretend we did not eat, did not consume calories, can ignore digestive processes that are beyond our view. If we do not grow or hunt our own food, we can believe that death is not part of life. If we do not acknowledge relationship, we can feign autonomy or dominance. Distraction supports destruction of Earth. Digestive and integrative functions in the body need safety and rest. Before eating, we might consider a silent moment for honoring the food—where it came from and all the people involved in its growth and preparation who care for

Spring Treats

Today I am baking Mother's Orange-Rhubarb Cream Pie, a seasonal favorite. The hardy and tasty red rhubarb stalks (and their giant poisonous leaves) are pushing up through the ground. The recipe is in my mother's handwriting and smudged from years of use. She so loved to bake and taught her daughters well: each step is carefully described, including how to shape a "high fluted rim" on the pie crust. Having made this pie for many spring gatherings, I greet memories surfacing as I spread the ingredients on our kitchen counter: eggs, softened butter, rhubarb, orange juice, pecans, and honey.

In the next hour, while the pie is baking and the scent permeates our kitchen, I'll call my friend Harriet Brickman to reminisce about her family's annual "May Day" parties with Maypole dancing, food tables in the field with piles of strawberries, local steamed asparagus with Hollandaise sauce, sliced country ham, plus all of our various pies and desserts—and children and friends. Delicious on all fronts. Off I go, to slice the stalks and dive into making, feeling my mother's hands and voice advising.

Mother's orange rhubarb cream pie
Pie and photograph by Jane Crosen

Dad's Lessons

My father had been a short-order cook on the railroad, putting off college to support his mother and sisters when his father disappeared—"deserting them"—during the Depression. Dad was a good breakfast chef. For a special Sunday treat, he would give our mother a break by making blueberry roll-ups—Danish pancakes wrapped around lots of blueberry sauce sprinkled with powdered sugar. The breakfast tradition, including candles and a fancy tablecloth, was passed on to my family. When my stepson Jonah was in a particularly rebellious high school phase, nothing I said or did seemed to have impact. Then he invited a girlfriend over and made her a fancy breakfast: lit the candles and cooked an elegant feast. Something from my family lineage made it through!

Dad's annual "trick" on April Fools' Day was to slip a small round piece of cardboard, cut from the side of a cereal box, into our dollar-sized pancakes, and watch: April Fool! I make them this morning for my husband, waiting for his surprise so I can snap a photo to email my sisters and niece. Sister Karen responds immediately, with an emoji at the end of her message, the facial expression telling all: "What fun! I picture the Illinois lake house kitchen, round table with us around it, and surprise and delight on faces when one of us gets a pancake different. Thanks for the refresh of family times and 'customs.' 😊 😄 —k"

Blueberries

Wild Maine blueberries decorate our morning oatmeal in California. This nutrient-packed breakfast links us to our home soil on the opposite coast. Small markers can make us feel connected, bringing a smile and a dose of appreciation to even a simple morning meal.

the land and farms that support our meals. Then we can enjoy the aromatic smells and flavors of food, the cultural matrix and profound satisfaction of digestion and interaction.

If food has been a challenging theme in one's life, re-establishing a relationship to food is important. Addressing food struggles including hunger, cravings, or addictive patterns is an important aspect of reconnecting with our own bodies. The gut and the brain are closely connected, impacting gastrointestinal function. Microbial gut-brain signaling affects homeostasis and ease—including mood and mental function.[4] Dissatisfaction and guilt around food and body image can undercut digestive efficiency. We may reject and refuse to absorb the nutrients that pass through us. Or we may consider food as a privilege to be tossed away or picked at according to whim. Working with difficult conditions like anorexia and bulimia, it may help to remember the digestive process not as something to control but as a cycling of powerful energy from the sun and nutrients from the earth through our bodies that eventually returns our digestive remains to the soil to nourish plants. Hunger has many dimensions. Survival is first and foremost, but love and connectivity are interwoven in the daily processes of creating, eating, and benefiting from our meals.

The body is intelligent; it has our best interests in mind. Working with, rather than against, this intelligence is one of the skills and the foundations of embodied communication. We can harness our best energies for what is important in our lives, like relationships, joy, and ritual celebrations. How do we do this? Often it is not the actual situation that challenges dining pleasure, it is our interpretation of all that is involved. Imagine if our meals were a time of restoration and connection. What might happen if we could use all of our nutrient intake toward meaningful interactions—co-creating a new way of relating to the present moment, to challenging situations, and to protecting the life-sustaining capacities of the Earth itself?

Daily dining and seasonal rituals are essential components of well-being and cultural cohesion. They offer nourishment and nurturance to self and community, and connection to larger patterns of life. There are so many ways that food impacts your living and communicating, from the nutrients you choose, to the pleasures or challenges of getting and sharing meals, to your cultural and historical embeddedness. Food may be easily available or financially or physically challenging to obtain; it may link you to others or separate you from family traditions. Knowing that you are constructing your body, your moods, and your interactions through food, you can be intentional about your choices. Enjoyment and embodied communication are part of the ongoing process of nurturing yourself with food.

Connections

Today the invitation is to slow down each time you eat to see, smell, and taste. Invite a few new flavors and textures into your meals or snacks to refresh your connection to food. Pause before you begin eating to shift your body-mind from action to digestion. Allow your imagination and appreciation to include interconnectivity with those who grew, harvested, prepared, and delivered this food to your hands. Notice with whom and where you most like to dine. And don't be hard on yourself—it is all good; no judgment is intended. Eating is a multilayered aspect of each day as well as an essential and potentially joyful part of embodied living and communicating.

TO DO

ACTIVATING DIGESTIVE FIRE (QIGONG)
5 minutes

This forceful breath energizes the digestive tract; it also gives you time to absorb.

Begin in basic standing posture:

- Bend your knees and place your hands on your thighs, focusing attention on your belly.
- Breath pattern: Draw the abdomen sharply toward the front spine as you exhale "ha" through the mouth. Alternate each out-breath with an in-breath (out-breath is "ha," in-breath is silent and through the nose) five times.
- Return to basic standing posture, placing your warm hands on your abdomen. Allow five natural breaths, absorbing the energy you have stimulated and listening to your body's response.
- Repeat three times, activating and receiving.
- Finish by circling the digestive tract (large and small intestines) with your hands, offering a light massage: move your hands up the right side of the pelvic frame (hip joint to waist), across the waist, down the left side, and then circle your hands a few times around the belly button to stimulate interconnectivity with the intestines.
- Pause and absorb.

TO WRITE

FOOD STORIES
30 minutes

Taste connects to memory. Take a bite of a delicious flavorful food: smell its aroma, feel its texture, taste its particular alchemical attraction. What was your favorite food as a child, and what do you love now? French fries or fresh asparagus; warm chocolate chip cookies or spicy gumbo? Follow the rabbit hole of your memory and include scents, tastes, and context. Did you have a particular family ritual or birthday meal? Do you have experiences of food in different cultural contexts? Write several food stories and their impact on your current life.

Tomatoes Tomatoes Tomatoes

We are planting twelve varieties of tomatoes. A friend has put in thirty-two plants this weekend, including Yellow Brandywine—a golden yellow beefsteak tomato we ate together last August. The slices were so large they filled an entire plate with their juicy mellow flavor. Catalog descriptions remind me of a craft beer menu, detailing scent, texture, and robustness in passionate detail. Our varieties include: Pink Berkeley Tie-Dye (sweet and spicy while sporting a rosy blush), Japanese Trifele (meaty and pear-shaped with a nearly black base), organic Damsel (globe-shaped, "absolutely bursting with juice"), plus Sungreen Garden (bite-sized lime green fruit that "top the charts in sweetness!"). Surely there will be bruschetta for picnics celebrating the bounty of summer (combining chopped fresh tomatoes with garlic, basil, olive oil, vinegar) and many fat summer sandwiches with mozzarella cheese and fresh basil. Planting takes work; imagining the flavorful results engenders motivation.[5]

Shrimp po-boy and fries
Superior Seafood and Oyster Bar,
New Orleans
Photograph by Josh Keith (2020)

VI
SOMATIC
PRACTICES

Jennifer Nugent and Paul Matteson
in *another piece apart* (2018)
Photograph by Ben McKeown

Mind and Mindfulness

Mind is a process, not a thing. Although we often associate the word mind with brain, the mind is not confined to one organ or location. The human brain is part of the visible, tangible world of the body; the mind is part of the invisible world of thoughts, feelings, attitudes, beliefs, and imagination. Although we can consider the brain and nervous system as tissues that facilitate mental function, all the body systems are involved in orchestrating our actions and reactions. In many ways, our actions shape our mind; where we place our attention creates neural maps and emotional networks, molding our tissues, informing our worldview, and impacting our embodied communicative potential.

Mind is vast like the ocean. Thought patterns are like waves on the surface; there is so much more below. Thoughts keep rolling in, calm or wild, but larger forces inhabit the unfathomed depths. Mapping the mind is an ongoing process. As a natural system, mind can be seen as an interconnected web that creates a whole. Pull on any strand, and it impacts all the parts. The cosmological dimensions beyond our specific Earth-oriented sphere are implicated: the stars, moon, sun, and galaxies continually influence our bodies. An expansion of the universe some thirteen billion years ago called the Big Bang distributed the materials that are now part of our structure. Meteors strike our planet and change geological components, the climate, and the course of biological life. And the primordial water that comprises much of our bodies, including 70 percent of the brain and heart, is the same water that has been on Earth for 3.8 billion years. It takes a big mind to conceptualize this vast heritage we inhabit. Although we like to name and identify its aspects, we also practice spaciousness when we reflect on the human mind and its interconnections with the universe.

We can begin our investigation of mind by noticing what is outside us. Your individual mind is being affected by the room or the place where you are right now. You have a measurable electromagnetic field around your body and so does the chair where you are sitting, the cell phone in your pocket, and the sunshine or light source illuminating your view, along with the plants and animals and people around you. Tuning to this energy field can be a way of describing the "mind" of the room based on the vibrational frequencies being experienced and interpreted in your body. Their impact affects how you feel, what you perceive, and how you interact with others. The seasons, phases of the moon, and presence or absence of sunlight affect your moods and rhythms of sleep and activity through your senses, as do the moods of your colleagues and family members. We are more interconnected through Earth's resonant frequencies and the electromagnetic fields around people's hearts than we recognize.

It might be correct to say that there are three times: a present of past things, a present of present things, and a present of future things. Some such different times do exist in the mind, but nowhere else that I can see. The present of past things is the memory; the present of present things is direct perception; and the present of future things is expectations. — Saint Augustine, from Confessions *(397–400 CE)*

Encountering Meditation

In my experience, meditation allows the practitioner to become aware of mental patterns—aware of awareness itself. My first meditation retreat with S. N. Goenka was later known by event hosts as the "retreat from hell." The gathering was on Cape Cod in Massachusetts in 1982, and it poured rain every day of ten. Bagworms dropped from trees and crawled into our tents; squirming was constant, but we were supposed to sit still, in silence, and *not* move. For ten long days, we focused on breath and body scanning. Sitting, sitting, sitting (not scratching), observing our minds. (The one calming presence was a giant tortoise resting along the worn path to the main tent; staring into the dark pools of her eyes, I relaxed.)

In the evening discourse, Goenka reflected back our thoughts from the day. Day one: "You are 'checking each other out'—who is here, how are they coping—instead of following your breath." (Accurate.) Day two: "You are thinking, 'This would be so great for my partner, my mother . . .'" Day three: "'How do I escape!'" And so on. By day ten we recognized our habitual patterns of thought as distractions, not-so-personal escapes from the present moment. "Thinking" is just a tiny fragment of mind. Watching thoughts, emotions, sensations, and intuitions pass by, like clouds in the sky, without attachment gives choice. This was where I learned that there's more to mind than constant chatter. I've participated in several less dramatic retreats since that time as well as continued daily practice. Always there is a clearing, calming, clarity amidst the challenge of staying present to the mysteries of mind.

As we move our attention inside the body, we can frame our understanding by differentiating and acknowledging the interactions between three aspects or centers of mind: the physical mind, emotional mind, and thinking mind. Science now confirms what many healers and spiritual practitioners have described for millennia: our gut, heart, and brain all have the neurological capabilities to direct our actions and shape our lives.[1] This overview of mind allows us to inhabit our wholeness by recognizing that the nervous system and endocrine system are interwoven, governing and communicating with all our body organs, tissues, and cells as well as with the world around us. Interconnectedness, or coherence, is our basic nature: the key to vital health and effective embodied communication.

PHYSICAL MIND

To locate the physical mind, place one hand on your belly and one on your low back. The space between your hands is the location of powerful kinesthetic, gut, and sexual intelligence. You can breathe into this space to stimulate its vital contents. The physical center of gravity in your body is located behind your navel (belly button), at the front of the spine in standing posture (the fourth or fifth lumbar vertebra in most bodies); powerful musculature is organized around the front, back, and sides of this region for efficient multidirectional movement and is supported by the breathing diaphragm above and the pelvic diaphragm below.[2]

The navel is the marker for your original source of life, the attachment site for the umbilical cord that connected you to your mother for physical nourishment in the womb. This also links you to the prenatal sexual energy that merged sperm and egg from your parents, creating your life.[3] The belly remains an important location for digestion, reproduction, and kinesthetic movement efficiency. The digestive organs (small and large intestines) fill this belly region; the lining of the digestive tract houses the enteric nervous system or your "brain in the gut," which consists of two hundred million to six hundred million neurons (similar to the number of neurons in the spinal cord). Creativity and gut-level intuition fuel and activate both grounded stability and passionate engagement that manifest from this powerful physical center in many forms of interaction.

Described through the nervous system, this belly-mind is linked to the brainstem as part of lower brain functioning, spatially below and evolutionarily earlier than the cortex. This quieter, older mind controls the vital organs, operating mostly unconsciously to support essential functions (such as heart rate, breath, immunity, digestion, and reproduction). Both parasympathetic and sympathetic nerve ganglia activate the belly-brain, influencing effective digestion and reproduction. For example: sexuality involves both receptivity (parasympathetic activation) and expressivity (sympathetic activation). Parasympathetic nerves in particular are located in front of the cranial/cervical and the lumbar/sacral segments of the spine.

The parasympathetic vagus nerve, the longest of the twelve cranial nerves, meanders throughout the torso and is an important communicator in stress, enabling the brain to keep track of the organs' actions. There are, however, more neurons going *to* the brain from the digestive tract (via the vagus nerve) than *from* the brain; the "gut brain" can override conscious thought.[4] Focusing attention on our foundational, kinesthetic, and gut intelligences, we see that these essential components of our physical brain allow us to function efficiently and effectively throughout our day, in conversation with the heart and head but also independent from conscious awareness.

A high level of kinesthetic intelligence through the physical mind allows for a feeling of groundedness, along with easeful coordination in the body. We likely know people for whom the body offers a balanced and stabilizing way of knowing and interacting with the world. We all have this life-enhancing capacity for "at homeness" to some extent, and it can be cultivated to increase skill levels and ease. Negative qualities of the physical mind include over-focus on appearance, problems with over- or under-eating, overindulgence or repression in sexual activity, and sexual predation or violence. Physical energy and intelligence fuel our whole body-mind and can be channeled productively or destructively according to intention and capacity for restraint. We know more than we can be consciously aware of at any moment; we rely on the belly-mind for survival reflexes, physical efficiency, and centered energy.

Mindfulness practices often begin by "taking our seat." This involves choosing a consistent place and position to root our thinking mind through embodied stability, creating a guidepost for concentration and awareness. In locating our base—pelvis, feet, or back in relation to cushion, chair, ground—and stabilizing our body, we create a responsive container to notice subtle energies of mind.[5]

EMOTIONAL MIND

The emotional mind can be stimulated by moving your hands to the heart region of your chest. This physical location of the emotional mind links to the much-studied emotional intelligence in the body. Encompassing the lungs and tissues around the heart organ itself, this region of the body responds to and initiates connection at emotional levels, affecting vital functions such as heart rate, breath rate, and blood pressure. Significant research in neurocardiology shows that the heart is much more than a "pump"; it has both neurological and endocrine functions as well as circulating the blood. Characterized as a "heart-brain," it has an intrinsic neural circuitry that enables it to act independently of the cranial brain to learn, remember, make decisions, and even feel and sense. The heart also has the largest electromagnetic field of the body: it connects us to each other and also to the Earth's magnetic field, enhancing our intuitive intelligence and energy coherence.[6] The lungs, which surround the heart and massage it with each breath, are important organs for both absorption and elimination; they take in life-sustaining oxygen and remove carbon dioxide as they support the heart.

At the level of the nervous system, emotional intelligence connects to the relational limbic brain that surrounds the brainstem. From a

Changing Your Mind
In my thirties, "changing my mind" was such a theme that I made a dance about it. We jumped up, fell down, ran this way and that way, danced this gesture, then changed to that gesture, pulled and pushed each other around in a mishmash of constant change to driving music. That was my state of mind. My underlying questions: "Whose voice am I listening to—yours, mine, my mother's, my father's, my critic's, my therapist's? Where IS mind?" It wasn't a very good dance, but it dispelled excess energy.

Nordic Mood

When I was growing up, a silent dark mood would take me over when my emotions were challenged. Brooding was a strategy I had both learned and inherited to keep others away, creating an energy field that announced: "Leave me alone!" Negativity is infectious, seeping out toward others. At one point, early in my marriage, I got so stuck in a dark cloud that I stomped out of our Vermont kitchen and headed off across cow pastures and open fields, through our neighbor's sugar maple stands, and past exposed bedrock, stomping, stomping, wearing down the energy inside. My husband was following me, unchallenged by my display of distance. When I turned to confront him, he said, "What's up?" (As if it wasn't obvious!) Direct communication changed my mood, lifting a curtain. My well-worn strategy was interrupted. It never really worked again.

physiological perspective, scientists looked for many years for the one structure responsible for all emotions in the human body. Eventually, they understood that it is the networking between various physical structures and our interpretation of the body's response to these processes that comprises our emotional lives. The limbic system, located at the base of the cortex, has been identified as largely responsible for processing emotions including memory and learning and as providing a foundation for all aspects of thought. It is particularly developed in mammals and primates to support bonding in live births and long parenting cycles.

A high level of emotional intelligence reflects a "feelingful," compassionate heart; empathetic awareness; and a capacity for emotional presence and decision making. Although anyone can cultivate heartfelt responsiveness, individuals with strong emotional affinity and elevated mood are often described as "warm-hearted"; they demonstrate a strong sense of values and commitment in relation to other people and belief systems. Diminished emotional intelligence can lead to an excess or lack of emotionality, dwelling in self-absorbed emotional dramas, replaying vengeful thoughts, or committing actions based on perceived emotional interactions or rejection. Hurt feelings in a relationship can result in shutting down all connections, a coldness or complete "change of heart" rather than resilience.

Many mindfulness practices begin with the breath since it is always present as an anchor to awareness. Building on a stable physical base, we focus on a compassionate but discerning heart. This often involves clearing past emotional tensions, blocks, and restrictions and channeling positive energy from nature toward our heart to support emotional clarity and heartfelt connection to the world.

THINKING MIND AND CONSCIOUSNESS

The thinking mind is located in the head. Place your warm palms on or around your skull, or tap it lightly to emphasize this connection. Many people associate the self with patterns of thinking, identifying with memories, dreams, expectations, and imagination. The thinking mind has been described as the element or complex of elements in an individual that perceives, thinks, initiates movement or stillness, wills, and—especially—reasons. The thinking mind develops and alters throughout our lifetimes, reflecting life lessons, stages of maturity, and changing conditions.

At the level of the nervous system, the brain houses both conscious and unconscious processes. The cerebral cortex is the wrinkled outer top part of the brain. The neocortex is the more newly evolved and largest part of the cerebral cortex (about 90 percent), with six neural layers that govern "higher functions" such as sensory perception, motor commands, spatial reasoning, conscious thought, and language. The prefrontal cortex (part of the neocortex) is located at the front of the brain and is involved in "executive functions" that help us navigate space, such as planning, goals, and actions. The back of the brain (hindbrain) supports more autonomic internal functions like breath, heart rate, and digestion that lead to inner awareness. With its multilayered and integrated complexity, the brain uses a tremendous amount of energy from oxygen and glucose; excessive thinking can drain other systems of essential nutrients, including the bones, heart, and reproductive and digestive organs.

Elevated thinking-mind functions include insight, spirituality, and creative visions that enhance life. Less refined mental activities have us rolling in repetitive thoughts, lacking clear judgment, playing negative mind games, and using words to distort or disguise truth. Although six emotions are considered by researchers to be universal—anger, disgust, fear, happiness, sadness, and surprise—it is important to note that the thinking mind tends toward the negative. We notice and prioritize what's wrong as a survival mechanism, scanning for potential threats and dangers. Mental processes can dwell in this cycle of negativity, operating on assumptions, projections, and thwarted expectations and believing that every negative thought is real. Agitated thinking can magnify stress beyond what is useful, flooding the body with an excess of activating hormones, such as adrenaline and cortisol, beyond what is necessary in our daily lives. This drains our energy resources. Mindfulness practices let us notice inner and outer stimuli without immediately reacting to them, creating an essential pause for enhanced discernment.

COHERENCE

Managing our thoughts is part of well-being. The popular saying, "Where our thoughts go, energy flows" reflects this attribute. Because we amplify what we dwell upon—by feeding it energy and attention—monitoring our thought patterns has significant impact. Balance or homeostasis requires circulating our attention and energy between these three embodied centers of the belly, heart, and brain without creating hierarchies or areas of avoidance. If the thinking mind is dominant, we can bring balance through more awareness of emotional or physical intelligence—heart- or belly-mind. If we are driven by physical desires or addictions, bringing attention upward to the heart and the thinking intelligence can offer support and elevated focus, qualities that are linked to these higher centers. Timing has impact. Sometimes we move too fast for our heart; feelings are slower than thoughts. Taking time to include all the centers allows more ease and fuller expressivity. Sometimes we neglect our physical intelligence, sacrificing the health of our body for outer appearance or to accomplish an external goal.

Coherence involves listening to and cycling between our multidimensional intelligences in connection to the people and places we love. In some meditation practices, we scan the whole body, circulating our awareness without attachment or hierarchy. Like the oxygen from our breath moving to every cell in the body through the bloodstream, awareness is local and global, cellular and whole-body. Meditation encourages returning to quietude, finding the inner balance that allows discernment during interactions.

Talking yourself out of healthy behaviors is all too easy; it is called

Stop Now!
"If you can't stop, run the other way." In his *New Yorker* article "The Art of Dying," Peter Schjeldahl is talking about his own experience of addiction. Even though we are essentially free spirits at birth, we can easily get lost in outer complexities, manipulated by other people's desires.[7]

Just Clap

Endings are important. Barbara Dilley, former Naropa University president and Merce Cunningham dancer, offers a simple and clear practice for ending one meditation session. She invites our group to count "one, two, three" and then clap together, tossing our hands up in the air to spread the good, focused energies into the world. It clears the air.

Internal practices require a return to full consciousness before re-engaging external demands. After a massage, a colleague backed her car into a tree. At the end of a ten-day meditation retreat, I rode home with a woman who drove our vehicle right across four lanes of traffic without even looking. We learn the necessity for ending practices with wakeful alertness, finding the appropriate state of mind for the next situation at hand.

rationalization. But you can also talk yourself into directing focused energy toward self-awareness and finding effective practices that support your best intentions. Energy is impactful in the body: you direct it as you will. The same energy can be used to help or harm someone, support or derail a project. Taking responsibility for your thought patterns is a first step in self-managing stress and inviting a more healthful, joyful life. You can ground down (connecting to earth), stand up (connecting to sky), and circulate energies throughout your body to invite full embodiment. Context plays an important role: How can your best, most coherent self be amplified or diminished in different life contexts? What choices do you have, and are there practices that can help increase and sustain a sense of integration, balance, and ease?

Mindfulness involves bringing the fullness of your attention to the present moment. As we acknowledge the need for managing our large brains, vulnerable hearts, and powerful physical impulses to moderate and modulate personal behavior, there are basic ethical underpinnings that support trust and ease in the process. Shared principles such as non-harming (to self and to others) and honesty (within oneself and one's community) create a firm foundation for receiving the benefits of impactful training practices. Linking personal well-being with social conscience and engagement involves discerning between the sensations and intentions that drive the actions of the physical, emotional, and mental aspects of mind.

MINDFULNESS LINEAGES

Reflecting on the long historical ancestry of mindfulness, we can consider its ancient roots as well as its contemporary, and largely secularized, manifestations. That genealogy now includes a recognition of nineteenth- and twentieth-century Asian Buddhists who reformed and taught mindfulness practices, including S. N. Goenka, who is referenced in the anecdotes of this text. American medical professionals like Jon Kabat-Zinn, founder in 1979 of the Mindfulness-Based Stress Reduction Clinic in the U.S., have translated practices to benefit health and healing. The commodification of mindfulness—utilizing meditation techniques without foundational ethical and social justice components—can be problematic. Scholars have articulated the unfortunate silencing of Asian and Asian American Buddhists in the teaching of mindfulness and the prioritization of the science of mindfulness over its historical and cultural lineage and nuance in relation to Buddhism.[8] As we cultivate positive qualities and clarify intentions in our contemporary embodied mindfulness practices, we can move into the present moment with awareness of the in-depth heritage informing present-day views.

Every culture has approaches to personal and social well-being for directing the propensities of an overactive brain, straying heart, and dimensional body. By valuing multiple perspectives, we can nourish the whole body-mind. As we access embodied intelligence through mindful living, we support effective communication in both easeful and challenging times.

Connections

Your inquiry today is to notice habit patterns of mind. Take a few mindfulness breaks: sit quietly and listen to your body-mind in depth. Clear the slate. Let go of obsessive thinking and doing if that is your usual mode, and practice embodied stillness. Rather than pondering past events or future plans and concerns, let your breath and the sensations throughout your body guide you back to present-moment awareness. Then as you move through your day and communicative interactions, you might be more present, less reactive, and more discerning of the best choices. In stressful situations, you will be prepared and "rehearsed" for less reactivity and more resilience.

TO DO

SEATED MEDITATION
5 minutes

There are many forms of meditation. This basic practice lets you notice the patterns of your thinking mind.

Seated, on a firm chair or bench, spine elongated:

- Rock forward and backward on your "sitz" bones, the protuberances at the base of your pelvis; shift your weight slightly forward (on the front of the sitz bones) to establish stability.
- Relax your hands on your thighs, palms downward for more containment, palms upward to release unwanted energies.
- Soften your gaze and turn your awareness inward.
- Focus on the breath, any sensations of breathing. When your mind wanders, which it will, bring it back to awareness of the breath.
- Your witnessing mind might notice sensations, thoughts, emotions, intuitions, or other stimulations of mental activity. Let these processes pass like clouds in the sky, staying attentive to the breath.
- For example, if sensations occur that are strong, you might just say "sensation" in your thinking mind, and return your awareness to the breath; do the same with other impulses. You can greet them and return your awareness to the breath.
- Being present without reacting does not imply inaction in life; we are practicing less reactivity in order to make space for more intentional responsiveness.
- Open your eyes, linking outside with inside.
- Take three deep breaths and find an ending to your practice.

Seated alignment in meditation is slightly forward on the sitz bones, the "feet" of your pelvis. This sends energy up the front of the spine. You may need several cushions to make this possible if you are seated on the floor; the knees should be lower than the hips so energy drains downward into the legs and the hips are relaxed with no strain.

What Matters

Lee Holden reminds us in qigong class that you really can't control other people's behavior, and often outside circumstances are beyond your sphere of influence—like being stuck in traffic. The only place you have actual impact is within your own mind and energy system. That's where real change is possible.

TO WRITE

OPPOSITE VOICE

5 minutes

> *Finding range in your writer's voice helps you access the multi-voiced diversity within.* Write an anecdote (short story, with beginning, middle, and end) using a writer's voice that is opposite from your own. Push the boundaries; what would you never imagine yourself saying? Let emotion ride out on your words through your opposite-voice story. Read aloud to yourself or a friend, embodying that voice.

PERFORMANCE TEXT

Awakening Grace—Mind[9]

Mind, it's vast, like the ocean,
unfathomed depths.
That mind, it can take you here and it can take you there.
It's a process not a thing,
and, it's always moving.

My friend Rich Wolfson, a physicist, says:
The wave is not the water.
It's just energy moving across the surface.
The water molecules remain largely in place,
until the wave hits the shore.

The mind,
it's not the brain.
It's just energy moving through the tissues.
The cells of the body
remain largely in place.

I am *reminded* of a time when I was six.
I was crawling under the bed, kicking
Your hands were on my white panties.
I didn't mind, really. We were just cousins playing.
But I hated you for that.
And I learned two things: Kicking works, sometimes.
And never trust anyone with those mindless, shifting eyes,
wounded.

I am *reminded* of a time in Paris.
I was a student. And I spent all my food money each week,
on one solo ballet class from a prima donna ballerina at the Paris Opera.
My teacher was curly-haired Joëlle Pinsard—a rebel.
She was teaching me the Bluebird Variation.
I was on demi point, and I started crying, fell totally apart.
I was hungry, too thin, out of my league.

She took me by the hand and put me on the back of her red
    Vespa motor-scooter,
and we drove around and around the Arc de Triumph, FAST.
I was . . . sobbing, my arms clutching her waist.
And then my sobs turned to laughter.
I put both heels firmly on the ground
and said goodbye to the prima donna ballerina in me.

I am *reminded* of the last time I heard your voice.
I was walking down a long hallway
and you called out "goodbye!"
I was late, my mind already racing ahead of me to my sister
    waiting in the car,
the plane, the flight.
It was the last time you asked me for something, and I turned away.

Clearing the mind, it's a daily practice,
making space for what's possible now.

Day 27

Body-Mind Centering®
Introduction by Bonnie Bainbridge Cohen[1]

Embodiment emerges from our cells' awareness of themselves. It is a direct experience. — Bonnie Bainbridge Cohen, from Sensing, Feeling, and Action

The Body-Mind Centering® (BMC^{SM}) approach to embodied movement and consciousness is an ongoing, experiential journey into the alive and changing territory of the body. The explorer is the mind—our thoughts, feelings, energy, soul, and spirit. Through this journey we are led to an understanding of how the mind is expressed through the body in movement.

There is something in nature that forms patterns. We, as part of nature, also form patterns. The mind is like the wind, and the body is like the sand; if you want to know how the wind is blowing, you can look at the sand. Our body moves as our mind moves. Changes in movement qualities indicate that the mind has shifted focus in the body. Movement can not only be a way to observe the expressions of the mind through the body, but it can also be a way to affect changes in the body-mind relationship. In Body-Mind Centering, "centering" is a process of balancing, not a place of arrival. This balancing is based on dialogue, and the dialogue is based on experience.

An important aspect of our journey in Body-Mind Centering is discovering the relationship between the smallest level of activity within the body and the largest movement the body is capable of making by aligning internal cellular movement with external expression of movement through space. This involves identifying, articulating, differentiating, and integrating the various tissues within the body; discovering the qualities they contribute to one's movement; how they have evolved in one's developmental process; and the role they play in the expression of mind.

The finer this alignment, the more efficiently we can function to accomplish our intentions. However, alignment itself is not a goal. It is a continual dialogue between awareness and action—becoming aware of the relationships that exist throughout our body-mind and acting from that awareness. This alignment creates a state of knowing. There are many ways of working toward this alignment, such as through touch, movement, visualization, somatization, voice, art, music, meditation, verbal dialogue, or open awareness.[2]

After many years of pursuing this path of inquiry with students, clients, and colleagues, I founded The School for Body-Mind Centering in 1973 as a means to formalize and articulate the ongoing research and as a vehicle for the continued exchange of information and discovery. At the School, techniques, applications, and principles are discovered and used in many ways. Some people practice the techniques, and then the principles emerge out of their own experience. Others focus on the principles, and as they apply them in their lives, they develop the techniques. But the techniques and principles themselves are not the material—it is the awareness and understanding of how and when to use them and how to invent one's own. The important thing is for each person to learn how it is that they learn, to trust their own intuition, and to be open to the unique styles of others.

In Body-Mind Centering we are the material, our bodies and minds the medium of our exploration. The research is experiential, as is the material. We are each the study, the student, the teacher. Out of this research, we are developing an empirical science—observing, contrasting, corroborating, and recording our experiences of embodying all of the body systems and the stages of human development.

For this science to have emerged, it has been essential to have many people involved in its exploration. Over the past forty-eight years, thousands of people have participated in the study and development of BMC, some briefly and a few for almost fifty years. The large number of people who have studied at the School have given the work its breadth. The people who have consistently collaborated with me over many decades have been essential in giving the work its depth. Their contributions to this body of work cannot be overemphasized. Together we have filtered through our uniqueness toward a common experience which embraces all of our differences. The BMC principles are drawn from this collective experience. The universal has emerged out of the specific just as the specific has emerged out of the universal. This is part of the nature of the work. As we go from cellular experience to that of the body systems, to personal relationships, to family, to society, and eventually to culture and world community, we are always looking at how these principles travel along a continuum.[3]

In Body-Mind Centering we use the maps of Western medicine and science—anatomy, physiology, kinesiology, and others—but BMC is influenced by the philosophies of the East as well. BMC comes out of a time where East and West are merging, so we work with the concept of dualities blending, rather than sets of opposites conflicting. We constantly look at relationships to recognize how opposite qualities modulate each other. Though we use Western anatomical terminology and mapping, we add meaning to these terms through our experiences. When we talk about blood or lymph or any physical substances, we are not only talking about substances but about states of consciousness and processes inherent within them. We relate our experiences to these maps, but the maps are not the experience.

The study of Body-Mind Centering includes both the cognitive and experiential learning of the body systems—skeleton, ligaments, muscles, fascia, fat, skin, organs, endocrine glands, nerves, fluids; breathing and vocalization; the senses and the dynamics of perception; developmental movement (both human infant development and human evolutionary progression through the animal kingdom); and psychophysical integration. As a set of principles and as an approach to movement, touch, and learning, BMC is currently being applied by people involved in many areas of interest, such as dance and movement arts; bodywork; physical, occupational, movement, dance, and speech therapies; psychotherapy; medicine; child development; education; vocal, musical, and visual arts; meditation; yoga; athletics; martial arts; and other body-mind disciplines.

While the basic material of Body-Mind Centering was well established

Bonnie Bainbridge Cohen
1987 Body-Mind Centering® Workshop,
Northampton, Massachusetts
Photograph by Robert Tobey

by 1982, the principles continue to be elaborated on and refined, with changes made as new viewpoints arise. As in any journey, what we perceive is influenced by what we have already experienced and therefore antici-pate. Consequently, BMC concepts are outgrowths of the personal histories, education, and experiences of the large number of individuals who have engaged in this exploration. As we analyze our experiences, the challenge is to not be confined by what we have already learned and to continually allow our discoveries to pass into our unconscious and to approach each moment with trust and innocence.

—BBC

Connections

Today your focus is on your organs—the vital core of your body. Use this opportunity to practice new perspectives from the Body-Mind Center-ing approach. Rather than looking for what's wrong in your body, open to moments in the day when you feel the most balanced and energized in the interiority and depth of your vital core. Specifically, notice your heart, lungs, and digestive organs. Of course, there will be times during inter-actions when your heart speeds up, your breath gets shallow, or your gut tenses. Yet, as you inhabit your body with attention and respect, you can feel the wholeness and integrity of communicating within yourself as you attend to the states of consciousness and the processes inherent within your organs. Radiance is the result: there is a brightness, brilliance, and luminosity that occurs as you bring love, generosity, and attention to your visceral body and also to your broader community.

Meeting Body-Mind Centering: An Embodied Approach to Movement, Body, and Consciousness

Body-Mind Centering became part of my life in 1978 when I met Bonnie Bainbridge Cohen in Amherst, Massachusetts. I was directing the dance department at Mount Holyoke College and took a student with a knee injury to see her for assistance. Bonnie said something to the effect that if you have surgery, you interrupt the integrity of the evolutionary line; you are starting with now. Through the needs of this student, I opened the doorway to my own learning—a new approach to my academic training in anatomy and kinesiology. I had studied anatomy seriously with my primary mentor, Dr. John M. Wilson at the University of Utah, and taught for a decade. But while working with Bonnie, you get to know yourself from the inside out. We explored through every system and tissue part by part, including embryological origins, evolutionary reflexes, and developmental patterns. "The pelvis" became "my pelvis," as I touched and encountered its unique curves, angles of force, and embodied memories. In Body-Mind Centering you move between the worlds of "learning about" and "learning from," respecting the inner intelligence of the human form. 👁

Samplings

In the online workshop series *Exploring the Embodiment of Cellular Consciousness Through Movement*, Bonnie offered a free introductory class. The first week, there were three thousand Zoom participants from seventy-seven countries including Iran, Iraq, South Africa, Kuwait, Egypt, and Lebanon . . . places that she'd never traveled to teach. It seems this virtual medium was a way to reach beyond the familiar global group. She stated: "Embodiment practices are now a much larger international community than we knew personally from traveling, teaching, and writing." There is a graciousness in hosting virtual sessions without rushing or over-exerting to communicate through the screen. She began with a pause, quieting into beginning.[4] From the class:

"We are moving through the whole body even when we're just moving the finger."

"As we live more in a community in our body, we know the muscles are not working on their own: they are expressing in the liver, the heart, every tissue. Opening ourself to our cells may help us live more communally in the collective world."

"By cellular experience, I mean initiation and awareness directed locally by the cells. Cells learn through new experience and tap potential patterns of behavior stored in themselves and in the nervous system. This dialogue between present cellular and past nervous system experience is what I call 'learning.'"

When Bonnie encourages us to move our heads, she adds: "Discover the cells moving your head, rather than just your brain saying to move."

"Reduce your effort, but open the possibility."

"Luxuriate in the gratitude that you are breathing. Luxuriate in the path of ease."

"Feel your vulnerability as a strength."

"Release the historical or cultural things that bind you: be free and be present now . . . Quit playing the old stories over and over again; get new stories. Choose the stories you want to live by: that's real courage."

"Learn to let go of what's inhibiting you. There are so many things that bind us; don't hold yourself back."

"Be alive with all your feelings; they aren't all good, but they are all real."

A question from the chat box: How do we recover from trauma? "There are patterns you developed for survival; life challenges took a strong response. First: respect the patterns, deeply respect them. Feel the feeling of the loss. Accept the stories, listen to the stories, respect the stories. Take time, don't force it, be patient."

"It's a practice, not a one-time thing. Breathing is a practice. Kindness is a practice. Opening is a prayer: opening to what's possible to receive and accept."

Movement and Mind
Describing her approach to embodied anatomy, Bonnie reminds us, "I think that all mind patternings are expressed in movement through the body. And that all physically moving patterns have a mind. That's what I work with."

TO DO

TOUCHING AND BEING TOUCHED—LOCATING THE ORGANS
10 minutes

Touch and movement can be helpful in locating and experiencing the internal organs. Begin with conscious touch, feeling through the tissue layers to skeletal landmarks and then bringing the warmth and energy of your hands toward the warmth and energy of the organ.

Stand with your hands on your belly:

- Pour your belly weight forward into your hands. Allow the organs to release toward gravity, and catch them in your hands. Jiggle the organs, feeling their weight, volume, and tone through the layers of abdominal muscles.
- Come back to your vertical plumb line and yield (don't pull) the organs back toward your front spine. Plump them out (we sacrifice so much for flatness); support with the abdominal sheath while keeping spacious in the organs.
- Place your hands on your *lungs*; they fill the rib basket front, side, and back, cradling the heart. Move with awareness of your lungs: the lungs are active, emptying and filling through the rhythmic movement of the diaphragm; imagine the alveoli gently bringing in the air.

**Workshop with Bonnie Bainbridge Cohen
Tokyo, Japan**
Photograph © 2016 Basha Ruth Cohen

- Place your hands over your *heart* area; the heart is located behind the protective breastbone (sternum) and slightly to the left. Move with awareness of your heart: the heart is a cardiac muscle, condensing and expanding (contracting and releasing) as it spirals and circulates the blood throughout your body.
- Place your hands on your *liver*, the largest and heaviest internal organ (wedge-shaped when seen from the side and about the size of an American football) filling the upper-right quadrant of the abdominal cavity between your lower ribs and pelvis. The smaller left lobe crosses the midline and helps to balance the weightiness of this organ (about three pounds).[5] Move with awareness of the dense liver that filters your blood.
- Place your hands on your *stomach*, on the left side of your body, partially under your lower ribs and between your ribs and pelvis. Move with awareness of your stomach: the stomach empties and fills, breaking down food to begin the process of digestion.
- Place your hands on your *kidneys*, located in the back of your abdomen, just under your back ribs, on each side of your spine. Move with awareness of your kidneys: the kidneys are fist-sized, bean-shaped organs that filter your blood and monitor fluids in your body; the right is slightly lower than the left due to the large liver.
- Place your hands on your lower belly, midway between belly button and pubis—a connecting point to both the ovaries (in female-identified people) and testes (in male-identified people).[6] Move with awareness of your *gonads*, the ovaries or testes: a base of identity and creativity.
- Orient to the weighted fullness of your organs. All organs have movement, condensing and expanding in spiral flow. Now that you have located these six organs, let them move you.

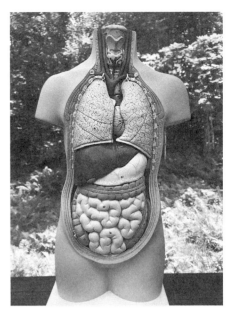

Organ model

TO WRITE
ORGAN EXPRESSIVITY
5 minutes

Organs are personal. Choose an organ that has meaning to you, and write from its voice: your heart, your lungs, or your stomach, for example. Enjoy listening to your organ and hearing the voice or voices that emerge. Powerful stories are stored in your vital core.

Day 28

The Discipline of Authentic Movement

Introduction by Janet Adler

The development of the inner witness is a good way to talk about the development of embodied consciousness. — Janet Adler, at the Somatics Festival 2019

Authentic Movement, compassionate witnessing of moving becoming conscious, is a transformative process grounded in the relationship between a mover and a witness. Different teachers of Authentic Movement offer their evolving perspectives in unique and diverse ways. The Discipline of Authentic Movement, a mystical practice centered in the development of embodied witness consciousness, is one way in this maturing and rich field of exploration.

For both mover and witness, attention is concentrated in the development of the inner witness, which is one way of understanding the development of consciousness. In this discipline the inner witness is initially externalized, embodied by a person who is called the outer witness. Another person, called the mover, embodies the moving self. After immersion in an ineffable field of movement, of stillness—accompanied by sensation, emotion, thought, or image—the mover returns to their witness, speaks what they know, and then listens to the witness speak their knowing. In this precious conversation, language ripens, bridging experience from body to consciousness. For each one present, commitment toward a practice of conscious speech is elemental for the development of this way of work.

In the early years of practice—inside this dyadic process named the ground form—the impassioned insistence of personal history, of personality itself, arises within the individual body. Within this safe crucible, a continuing practice evolves as movers become moving witnesses and moving witnesses become silent witnesses. Each role integrates enough, and when ready, guided by one's teacher, the great responsibility of becoming a speaking witness is consciously chosen—first with one mover, then with more. Practice deepens, and participation in a collective body of movers and witnesses is invited as everyone becomes an integral witness: one who empathically witnesses, listens, and responds by speaking consciously from multiple perspectives.

Being enfolded within these specific archetypal shapes—formed by humans longing to be truthful—strengthens each one for other ways of knowing. Intimate in the presence of each other, an emergence of energetic phenomenon, if or when it naturally occurs, is consciously received and welcomed. This transpersonal energy, which does not arise from personal narrative, becomes intuitive knowing. In moments of grace, one is an empty mover—an empty witness. Authentic compassion, compassion that cannot be willed, becomes the vibrant force within the web of embodied connection.

Janet Adler
Somatics Festival 2019,
Northampton, Massachusetts
Photograph by Mandy Kimm

Being seen, seeing oneself, and seeing another, movers and witnesses move closer to their light because of consciously embodied surrender into the vital energies of relationship with another being, with their inner life, with collective bodies, with that which is invisible. Through this mysterious, developmental but nonlinear process of enduring commitment, directly knowing a trust in themselves and in the discipline, individuals can journey from the experience of duality to unity consciousness.[1]

—JA

Connections

The invitation today is to enter an experience of moving from inner impulses. In Janet's work, the experience of moving in depth preceded her immersion in the Discipline of Authentic Movement. Close your eyes a few times during these next twenty-four hours, and let yourself move and be moved! The body knows at every moment what you most need for rest, recovery, expression, and connection. As you tune in and listen to these requests of your embodied self, you can recognize what is calling for your attention. Moving authentically is highly specific; it's not just "anything goes." As you engage deep listening, you'll know exactly where you want to be in a room, who you need to be near, when you need distance, and whether you are drawn to light or to darkness. Today, as you engage moving from this inner listening, you open to a vast terrain, appreciating your dimensionality as well as the possibilities of discernment and choice.

Moving and Being Moved in Authentic Movement

I met Janet Adler in 1979 at an art exhibit and asked if she might be interested in introducing her work in Authentic Movement to our (young) dance company. Dance training, in my heritage, involved teaching the body how to do things. Because I had excellent teachers from a young age, this required inner reflection: noticing the role of the mind on buoyancy and balance—for example, the physics of grounding through a plié to spring into the air, and the influence of thought patterns on motivation and intention to hold an arabesque. Dance also trained a deep respect for the threshold of entering and beginning a process. As you stepped through the doorway from daily life (the waiting room) to the studio, you left your petty, distracted self behind and entered a realm where movement was primary. The entranceway was a transitional space to change mind and attitude.

Authentic Movement with Janet in those early years of her inquiry reflected this heritage of dance training. We closed our eyes in the presence of her witnessing and focused inward. In this process, rather than directing the body, you allow inner impulses to move you—moving while being moved. It was surprising at the time to realize that another whole realm of movement possibility was right there waiting: personal, collective, and transpersonal. Mostly, I learned that we know so much more than we think we know. Practicing moving between the worlds of inner and outer awareness in the presence of a witness, you develop the capacity for discernment. My development of Authentic Movement has focused on its connections to artistic practice. The Discipline of Authentic Movement is Janet's more recent refinement of the process, with its own distinct training program for embodied consciousness as a mystical practice.[2] 👁

Somatics Festival 2019: Celebrating the Work of Janet Adler, Bonnie Bainbridge Cohen, and Nancy Stark Smith, and the 45-Year Heritage of *Contact Quarterly*

This four-day gathering was held in Northampton, Massachusetts. Here are some quotes from Janet's offerings:

"It all starts with the outer witness trying to see clearly. With that
    compassionate presence, the mover comes to see oneself more clearly."
"The light of the inner witness appearing within each mover."
"And how incredibly harsh and unkind the inner witness can be."
"And how we can grow a kinder inner witness."
"It is possible to be fully oneself in the presence of many others without
    betraying oneself."
"As the inner witness clarified, this way of work—The Discipline of Authentic
    Movement—for me is a mystical practice."[3]
"You know that gorgeous cross where the horizontal is about the mystery
    of the collective, and the vertical is the traveling within each of us, top

to bottom or earth up through. We all have our experiences of that inner journey."

About colleagues who are inquiring in this embodied way: "The word somatic didn't exist in the 1970s–1990s when I was in Northampton, Massachusetts, so when we began to know each other, we shared so many students; I felt so supported. 'Oh, finally, I am not wandering around with this word embodiment and this devotion to this phenomenon alone.' And now in the world: how many choices people have in their discerning how they want to explore their own experience of embodiment, and so many are offered with integrity and goodness."

For videos of panels, workshops, lectures, and historic slide shows documenting Janet's views, see: www.somatics2019.com/.

TO DO

EYES-CLOSED MOVING

10 minutes

> *Within the Discipline of Authentic Movement, there is an outer witness or witnesses. The following movement invitation offers an introduction to eyes-closed moving as an individual awareness practice. Read through this full set of instructions to create a safe container for your exploration.*

Begin in any position—right where you are. Close your eyes and bring your attention inward:

- Notice any impulses for movement that are coming from your body (not just thoughts about what to do).
- Allow yourself to follow your body's desire for movement: it might be a tiny micromovement or a big stretch. You can make a choice to attend to an impulse or let it pass.
- Follow your inner intelligence, letting your body be your guide.
- Explore self-witnessing; there's an inner flow that directs while you maintain safe boundaries.
- Bring your movement to a close; slowly open your eyes and take time to reorient.
- Resist judging; instead, value and contain your experience as you create a dialogue with your inner movement life. Writing, eyes-open moving, or drawing can be a useful transition.

Three rules: Open your eyes if you want to move fast or through space; follow your own impulse if you come in contact with someone/something else; and you don't have to do anything hard. If memories or uncomfortable sensations arise, you have choices: open your eyes slightly, or pause and change the shape/position of your body slightly to lessen the energy. Within that context, allow yourself to move and be moved.

TO WRITE

FIVE QUALITIES OF YOUR INNER WITNESS

10 minutes

Cultivating a supportive, non-judgmental, but discerning inner witness is part of Authentic Movement practice. Make a list of five qualities you would like your inner witness to have, supporting you on your investigation. Perhaps your list includes kindness or a sense of humor. Let yourself be surprised by what you choose as essential. Then write about those qualities and why they are important to you. The goal, of course, is to inhabit those qualities as you witness yourself writing—replacing the inner critic with qualities that bring out your most authentic, original voice.

Qigong

With Lee Holden, founder of Holden Qigong

Qigong, or "skillful energy," is a broad term for multiple contemporary wellness practices with roots in ancient China focused on moving energy through the body. Over two thousand different forms of qigong are practiced worldwide. Within those forms, more than two thousand movement sequences are done standing, seated, or lying down—indoors or in nature—to support health and longevity. Qigong is now a widely utilized movement practice globally, with people engaging daily on every continent and within a variety of social, cultural, and educational contexts.[1]

Qigong originated about four thousand years ago in China.[2] The earliest forms of qigong had various names, one of which was dao-yin, which can be interpreted as "leading and guiding the energy" and was linked to Taoist philosophy. Concurrently, yoga was developing in India, and some of the postures and principles overlap. Teachers and students traveled, shared findings, and returned to their homelands—developing practices with their own cultural and personal signatures, just like today. Early qigong forms were so effective at building energy and harnessing and moving qi (chi) for health and well-being that they became the foundation for a martial arts practice now known as tai chi. Linking breath, movement, and mental intention for optimal health and vitality, current Western approaches honor the ancient heritage of various qigong and tai chi lineages, while addressing contemporary health needs. Their foci include helping individuals and groups clear stress, recharge energy, and calm and focus

When your body relaxes, your mind relaxes. When the body is open, the mind is open.
— Lee Holden, from Qigong Workbook

Lee Holden, Swimming Dragon

Interview

From Lee Holden's "Tea with Lee" web conversations hosted by Ben Cox, 2020

What is the connection between Western and Eastern anatomical approaches?
"In Western science, we don't have an 'energy system'; we identify the circulatory system, respiratory system, cardiovascular system, nervous system, and so on. But each of these systems needs energy—the bio-electric energy field that keeps life flowing. In qigong, energy travels through meridian lines that are used effectively in acupuncture for balancing energy flow and also in hospitals in China to anesthetize the body."

How do we come back to the present moment when the mind starts to get stressed?
"Bring attention to anything that is happening in the moment. Where am I right now? I feel my back on the chair, I can see . . . , I can hear . . . Tune in without judging; be present. What happens when you stabilize present moment awareness? You feel safe, content, calm. When your mind starts to get stressed, ask yourself: How do I know I am here? You can observe (I see, I feel, I hear, I taste) without judgment. This is mindfulness in action, the skill set of working with your mind."

Why does qigong put so much focus on breathing?
"We breathe about 20,000 breaths a day; breath links unconscious and conscious aspects of our lives. When you change breath habits, you are working into the programming of your unconscious mind. To breathe slower and deeper sends the cue: 'Everything is okay, relax.'"

How can I sustain my practice?
"Cross-train your routines. Change with the seasons. Have different morning (sunrise) and evening (sunset) routines. Take a class or a workshop. Keep it fresh and fun and connected with the community."

(*continued on the next page*)

the mind. Significant research is now available linking qigong practices with medical well-being initiatives internationally that create the conditions for more effective engagement in life and work, including the self-management of stress levels.

A first step in qigong is becoming aware of subtle energy and the pathways it travels through the body as well as how to draw energy from the natural world around us. Qi or "life force energy" is like water: you do not want this essential energy to get stagnant; it needs to move, flowing and circulating through all the body systems. Although qi is invisible, we generally know when we have energy and when we feel drained or dull. One way we recognize invisible qi is by noticing its visible effects on the body through sensations such as heat, vibrations, pulsations, and tingling. Ask a friend how they feel, and a response might be: "I feel good, I have lots of energy." There's a sense of having enough vitality for life's demands. Or the opposite: "I feel drained and unmotivated." Sometimes you just can't get moving, reflecting low energy. By self-managing our energy flow, we can communicate more skillfully in different contexts rather than draining and wasting our energy reserves or depleting other people's energy with our moods.

Qigong is one of seven branches of traditional Chinese medicine along with acupuncture, herbal medicine, nutrition, massage therapy, feng shui, and cosmology. Chinese medicine is preventative in nature: you come to your health care practitioner or medical team to keep you healthy. This ancient form of "health care" is a kind of mind-body medicine with wellness at its core. Ancient qigong and Western medicine (which developed around two hundred years ago) are complementary. Western medicine is particularly effective at treating disease and catastrophic accidents and offering protective immunizations. By working together, Eastern preventative medicine helps you stay healthy, but if something interrupts optimal health, Western medicine is efficacious at addressing acute problems.

Over the millennia, qigong has branched into three main areas: medical qigong (using exercise and movement to stay healthy), martial arts style qigong (building power and strength in the body), and spiritual or meditation-focused qigong (cultivating spiritual connections and accessing one's life-purpose). The "to move" exercises utilized in this text are from qigong master and health practitioner Lee Holden and are drawn from the medical qigong tradition. As a principle-based practice, Holden Qigong offers anatomically accurate descriptions of exercises, fluidly linking Eastern practices with Western medical models. Activating energy, clearing blocks and tension, and flowing through sequences help energy move throughout the body as we connect with nature's cycles. With qigong, you keep your body moving at all ages and stages of life for health and longevity.

The name qigong can be confusing. First, we have no words in English that start with qi, so pronunciation is often a challenge. And there are multiple spellings in English, translated and transliterated from written Chinese characters (logograms) since the language is pictographic.[3]

The term qigong (pronounced chee gong) is a relatively new name, which came into usage in China in the twentieth century with the Communist Revolution. It was used to separate the physical practices of qigong from their religious and philosophical connections so they could continue to be practiced in the context of newly secularized communist China. Focusing on the principles of qigong helps bridge the gap in our understanding. Many of us may be more familiar with its underlying philosophy and the yin/yang symbol from China than we are with the name qigong.

In ancient Chinese philosophy, yin and yang reflect the concepts of dualism and complementarity. The yin/yang symbol shows a balance between two opposites, with a portion of the opposite element in each section, represented by a small dot in each half of the opposite color. The underlying principle of yin and yang is that all things exist as inseparable and complementary opposites and expressions. For example, female-male, dark-light, and old-young. The principle, dating from the third century BCE or even earlier, is a fundamental concept in Chinese philosophy and culture in general. At the fullness of each expression there is a seed of its complement expression. For example, at the fullness of a yin cycle (like winter solstice), there is the seed of a yang cycle (summer solstice), thus reinforcing their complementary natures.[4] Encouraging moderation in responses and tolerance of opposites, the symbol is still applicable and widely used centuries later. Rather than seeing opposites in conflict, we experience complementarity. Dark and light are both part of life, and often one changes into the other. For example, you might experience something as negative (spraining an ankle), and it may turn into a gift (you have time to write your book). Or what was good (a financial raise at work) may turn negative (too much work). The circle around the outer rim of the yin/yang symbol contains the whole, representing the cyclic nature of life.

As an ancient set of practices, qigong has a richly varied heritage—physical, philosophical, and theoretical—with multiple aspects of research and understanding that apply to our contemporary lives. Because qigong seeks to balance the foundational energies in the body, it is an excellent complement to embodied communication skills. Self-managing stress, respecting the energy and practices of other people and cultures, and conscious awareness of the energy you bring into a room or conversation are a few of the many benefits sustained practice can offer.

Connections

Experiencing energy through qigong is your focus on this day. Since energy is invisible and embodied sensations are not-always tangible, start with the simple qigong practice of shaking your body parts to activate energy flow. Include the five energy gateways leading to your vital core: shake through the joints in your arms, legs, and at the base of your skull. Then slow your movement down to notice sensations: you may feel tingling, heat, or trembling as manifestations of energy. Repeat this at various intervals in your day, noticing how energy moves through your body and

(continued)

What is contemporary Western qigong?
"Western approaches to qigong address common issues from our particular lifestyle, namely too much stress, and not enough energy. Qigong always included specific mental intentions to accompany each movement. The focus of this modern approach is to help people clear stress, recharge, relax, unwind, cultivate inner peace, and calm the mind. Contemporary practices celebrate the history of qigong while integrating elements of the practice to address specific challenges to our minds, bodies, hearts, and souls."

Can I practice on my own?
"Self-directed practice motivates you to remember. On days when you can't get to a class, or don't want to turn on an electronic device for virtual participation, it can be useful to have a sequence that you know well enough to do without an instructor. What do you remember? What feels good? Trust that your body remembers what it has learned; sometimes you are your best teacher."

Note from the Holden technology team:
"Technology can scatter qi. There are large amounts of electromagnetic pulses through electronics. It is not just the screen-time, it is the physical effect of touching the keypad. If you spend a lot of time on the computer, it is hard to put it down. Most of us get dopamine hits from feeling connected to others, even in this virtual realm. Sometimes you have to consciously have periods without screen time: take breaks with all wi-fi turned off. It's a perfect time to do some qigong."

Yin/Yang symbol
Image: iStock

Shuǐ (water), calligraphy by
Chungliang Al Huang
Courtesy of Livingtao.org

connects to nature and other living things and beings. You are "checking in" rather than "checking out." Consciously circulating energy between your head, heart, and belly throughout this practice can enhance your living and communicating by bringing more ease and flow.

Qigong and Tai Chi

I first experienced the lineage of qigong as the underpinning of tai chi energy flow with Chungliang Al Huang in 1971.[5] He offered classes and performed his unique combination of modern dance and tai chi (tai ji) at the University of Utah, where I was a graduate student in dance. I repeated his concise *Five Moving Forces* practice daily outdoors for two decades and joined him in 1986 for a residential retreat at the Lan Ting Institute in Fujian Province, China for intensive tai chi study. Arriving on narrow bamboo rafts poled by skilled boatsmen along the Wuyi river, a resilient stance was our introduction to embodying flow.

In 1996, qigong scholar and master teacher Kenneth Cohen introduced me to the standing practice of *Embracing a Tree* when we were both presenting at a Common Boundary Inner Ecology/Outer Ecology Conference in Washington, D.C. In this energy-building posture, workshop participants stood with arms forming an open circle in front of our chests, knees slightly bent, remaining "still" while he spoke insightfully about qigong theory and history for what seemed a very, very long time. If you were doing this standing posture correctly, qi moving through the body would keep your arms and legs from throbbing. (I was in pain.)

Yet in 2016, when I moved to Monterey, California, qigong master Lee Holden was teaching in nearby Santa Cruz, and my husband and I made weekly excursions to his classes for two years. Lee easefully and effectively integrated my desire for accurate cues around anatomy and physiology (Western science) with Eastern medical principles through a flowing movement form of qigong. I am now a certified Holden Qigong instructor and practice daily.

Qigong is easily applicable in one's life to clear stress, build energy, and increase flow. You don't need props or special clothing: you can do it anywhere at any time—indoors or outdoors—and you can immediately feel its effects on your mood and health. Although hard to pronounce for Westerners, qigong is deeply impactful and relevant for reducing stress and increasing energy. I am pleased to offer Holden Qigong within the context of this book as a resource for living and communicating with more ease and efficiency. ◉

TO DO

BAMBOO IN THE WIND (QIGONG)

1 minute

Qi (Chi: Breath), calligraphy
by Chungliang Al Huang
Courtesy of Livingtao.org

This standing exploration is a good way to end your practice, connecting with the plant world to cultivate replenishment, contentment, and presence.

Bring your feet close together and place your hands over your lower abdomen, eyes closed or with a soft focus:

- Take a few deep breaths into your belly.
- Relax into your legs. Feel like your body is bamboo swaying in the wind, gently rocking in an imaginary breeze.
- Have the sense of being rooted and grounded through your feet but flexible through your body as your head reaches to the sun.
- Reduce your effort; movement is happening all the time without your need to do anything.
- Rock and sway with your eyes closed for about one minute. This rocking motion is natural and will allow your body to unwind and come back to a place of feeling grounded and centered.
- Open your eyes slowly, staying aware of breath and ease.

TO WRITE

REFLECTING ON QIGONG

10 minutes

Change begins with awareness. Settling into the present moment through breath and sensation, write about your experience of qigong and the movement of chi in your body. Consider how a consistent practice might be useful in your life.

Caryn McHose
Arches National Park, Moab, Utah (2016)
Photograph by Kevin Frank

Body and Earth

With Caryn McHose, Body and Earth co-developer

Body and Earth connects body systems and Earth systems. Body is Earth; our bones, breath, and blood are the minerals, air, and water inside us—not separate, but same. This basic principle in the Body and Earth practice helps us recognize that interconnectedness is fact, not metaphor. Scientists have confirmed this knowledge by researching and measuring aspects of both the human body and Earth's body. Our physicists, geologists, biologists, and neuroscientists have proven what many sages have understood experientially through the eons: we are intricately interconnected with the world around us. The water you drank this morning is now inside you, part of the 70 percent of the body that is water. And that water has been on the planet for 3.8 billion years according to the geologic record. It is the same water that passed through the bodies of the Buddha, Cleopatra, Martin Luther King Jr., or any of the figures in history or in your present life who inspire your imagination.

Humans do not create movement; we are part of a moving universe. In the Body and Earth practice, we honor and appreciate the inherent intelligence of our moving, breathing bodies in relation to the places we inhabit. Rather than seeking control over our bodies as objects or machines or seeking dominance over Earth systems, we encourage creative dialogue—circulating awareness between inner and outer realms. In this body of work, nature is primary; everything is alive. Passionate inquiry is required to articulate in words how the differences and multiplicities of our relationships to the Earth are central to the ways we re-engage outward.[1]

The intention of this practice is to move more easefully between the various dimensions of our lives, to not get stuck in any one way of being in relationship with the world and all its creatures. As we consider ourselves to be participants in a moving universe, we acknowledge that we don't live on the Earth; we exist within the Earth's body, surrounded by a life-sustaining atmosphere, while standing on the moving, breathing skin of the soil.

Body and Earth training workshops circulate between five key "bodies of knowledge," exploring their practices experientially as well as intellectually. These categories (pods of investigation) include experiential anatomy, evolutionary movement, perception, Authentic Movement, and moving with nature—acknowledging spiritual dimensions of interconnectedness. Relational inquiries underpin the whole: moving between worlds through communicating with oneself, others, and the places we inhabit.

Change is inherent. Annually, we reflect on these areas of investigation and meet them as old friends but with new levels of understanding and

This body is always in an exchange, a reciprocity and rapport with the larger body of the earth itself. — David Abram, from the presentation "Between the Human Imagination and the Land's Wild Dreaming"

integration. Explorations in embodiment take focused time and commitment. Yet as experiences deepen, they are available spontaneously, instantaneously in our daily lives. As we value movement as a way of knowing ourselves and the world, we can recognize that all humans have a body—we each are a body—and there are 7.8 billion human bodies on the planet to feed and value. Any sense of separateness between the wellness of the human body and the wellness of the Earth body is simply, and dangerously, an illusion. Wholeness is something that lives through us as we inhabit inner and outer terrain. Cultivating mystery is an edge: standing with one foot in all we know and the other in all that we do not know, we sustain curiosity.

Connections

During your day today, connect with five Earth systems that are present outside and inside of your body: water, air, plants, animals, and minerals. At various times, look around the spaces where you are (indoors or outdoors) and notice where water is present, the quality of the air, the wood in the floor or comprising the book you are holding, the animals nearby, the gemstones you might be wearing, and the minerals/rocks/metals that are visible. Then do the same, noticing the water, air, plants, animals, and minerals inside you or on your body. Engaging the Body and Earth lens, you can connect to your aliveness, not as dominant species but as part of a living, breathing Earth. Taking time to notice this active engagement can change your day, calm your mind, and support your best self as you feel the depth of interconnectedness.

Susanna Recchia, Marloes Sands
Beach, Wales (2015)
*Body and Earth: Seven Web-Based
Somatic Excursions*
Still image from videography
by Scotty Hardwig

Body and Earth-ing

Caryn McHose and I became colleagues at Middlebury College in Vermont. She participated in hiring me in 1982, and we have worked together in various settings ever since. The story goes like this: I was considering leaving my professional life in a dance company to take a college position with financial security. My fear was losing my creative edge and artistic colleagues. Caryn was one of the dance faculty who interviewed me for the position of Assistant Professor of Dance; she listened over lunch and then drove me in her beige Volkswagen bug up to the glorious New Haven River near Lincoln, Vermont. We waded out to the middle and lay on a huge flat rock amid that roaring downhill flow, wordless. I thought "I can do this." For several years we explored together, tucked away in the basement squash courts of the old gymnasium, deepening our explorations in anatomy, voice, and movement supported by that lush Vermont land. We have taught together and separately since that time, always exchanging areas of inquiry and discoveries. You never know when someone or something will change your life, requiring something new of you. Caryn's intuitive approach has been central to my journey with embodiment, dancing in nature, and the development of this Body and Earth material, which continues to unfold. 👁

Interview with Caryn McHose by Donnella Wood at a Body and Earth Retreat in Holderness, New Hampshire, July 2019.[2]
For those who are less familiar with embodiment practices and somatic approaches, how do you begin?

"For those new to the practice, the first question is 'Where am I?' rather than 'What is going on?' Can I begin to notice that I'm supported by the ground, even if only for a second? Can I stay fresh to what's actually happening rather than what I think is happening?

"Allowing one's body to feel how the ground supports us is a foundational perception that interrupts old patterns and reduces effort. Noticing the space around us is the other part of the equation. Both ground and space provide information that is needed to be present to our current location, allowing us to adjust.

"Feeling the weight of our bones, the weight of our flesh, is information that lets our movement brain—as part of the tonic system—coordinate with more ease, reducing effort. Noticing a tree or something at a distance can lighten and elongate our bodies in relation to the space where we reside. These two factors, the weight of us and the space of us, are a return to inherent organismal intelligence, our coordinative birthright.

"People know they're supposed to be in their body. They want to have a practice. They know it's an important value. And then I notice them using the expression 'I've got to be grounded.' And the 'grounding' word is staying in an abstraction. That's a representational framework; it becomes a symbol. And then if I ask: 'How do you notice that?' there can be a deepening of sensation in the body. In that inquiry there's an interruption of habituated ways of moving, and a new moment of engaging. Coordination is refreshed.

"Once you empower someone to feel the difference between a word that is staying relatively abstract to a word that is being felt in the moment, you are in the realm of embodiment. Even in experiential anatomy, there's a journey between seeing a representation of a part of the body—like a drawing of the bones of the foot—and taking the time to feel and differentiate that body part in yourself."

What are some of the benefits of the work?
"When people start doing this work, I see a slowing down, a plumping out, and a warmth circulating. I see inherent stability and strength expressed without rigidity. I see people being able to feel what they haven't been able to feel, and then becoming more fluid with their emotional needs and creative offerings. There is a restoration of the capacity to resonate, empathize, and connect."

Can you describe some of the levels of abstraction that happen with language?
"We, as human beings, find words that abstract and re-present the lived experience. At any given moment, I can ask myself, is this wording referencing a lived, felt experience that can then be named as a sensation? For example: 'I am noticing a spreading or a softening as I feel the support of the ground' rather than 'I am grounded,' which is more abstract. Can I be aware of how quickly I make associations through images or other levels of conceptual thinking rather than staying with the direct experience of the sensations in my body? Through these embodiment practices, one can circulate between feeling, meaning-making, and communicating with others without losing the felt sense of the body.

"In the Body and Earth work, I try to support the inquiry around: 'Where am I?' This involves feeling ground and space and recognizing the sensations of the body in the representational process of speaking—finding wording from the body rather than about the body. Then, language can circulate within the lived experience."

TO DO

THRESHOLDS

15 minutes

Thresholds are connectors.

In a studio or outside: find a doorway that interests you. Walk across the threshold. Notice the approach, the passing, and the aftermath. Keep walking and explore:

- What's the threshold between walking and moving? Take time to explore.
- What's the threshold between moving and eyes-closed moving?
- What's the threshold between eyes-closed moving and eyes-open dancing?
- Between dancing and raucous dancing?
- Between raucous dancing and performing? (Consider "performing" as any fully engaged action with intention to communicate.)
- Between performing and meditating?
- Between meditating and walking?
- Repeat the sequence and notice what you find.
- To end, add the threshold between walking and stillness.

TO DO AND WRITE

PLACE SCAN

10 minutes

A place scan includes both outer and inner landscapes. Scanning through five aspects brings your awareness to place before writing.

Begin seated, eyes open:

- Bring your attention to soil and rock. Where is there soil in this room or place? If you are indoors, take time to consider the possibilities within this place and nearby: metals, glass, sand, and rock. Be sure to include your own body and other humans—your bones, the gems on jewelry, the metal in zippers or accessories.
- Bring your attention to air. Where is there air in this room or place? What can you smell, taste, feel, hear, or see concerning air? Consider what is happening with the seasons and with the sun and moon. Bring your attention to any aspect of air that you can perceive at this moment in time.
- Bring your attention to water. Where is there water in this room or place? Consider what is happening with the water cycle, the water table, the watershed. Remember your body is mostly water.
- Bring your attention to animals, our relatives on this Earth. Where are animals in this place? Notice the season, the foods

ENGLISH—FILIPINO—SPANISH

Walking—*Lakad*—*Caminando*

Moving—*Galaw*—*Moviendo*

Eyes-closed moving—*Totoong galaw—Movimiento natural autentico del interior*

Dancing—*Sayaw*—*Bailando*

Raucous dancing—*Malat na sayaw*—*Baile estridente*

Performing—*Pagsasaga-wa*—*Performando actuando* (*presencia escenica*)

Meditating—*Magnilau*—*Meditando*

Try speaking the thresholds sequence in your home language; how does it change your experience?

Passage (2019), sculpture by Herb Ferris
Vermont Institute of Natural Science
Quechee, Vermont
White cedar, steel, 18 ft.

you have eaten, the clothes you are wearing, the insects and tiny micro-organisms.

- Bring your attention to plants. Where are plants visible in this place? Even if there are no plants in sight, the floors may be wood, plants are being digested in your stomach, plant fibers may be in your clothing or books. Notice the plants around you that are providing oxygen sustaining your life.

- Pause in open attention. Try scanning around you with senses heightened toward body and earth through a quick scan of the soil, air, water, plants, and animals of the specific place where you are.

- Write about your experience. What is easy to notice; where are you hesitant or surprised; what memories or descriptive elements are evoked in this place where you are sitting?

VII
PERSONAL PROJECT

Your Chapter

Having a vision and moving it through to completion is a process. Create your own chapter to add to this book. Consider what is missing or intriguing from this text that you would like to explore in more depth. What ideas, images, and experiences link to your lifework and creative imagination? What somatic excursions would help someone else to explore your topic? Be realistic about your timeframe. The task is to begin. Don't just think about it, envision your project and bring it into fruition; share it through embodied communication.

You begin with a spark of enthusiasm for a topic—a vision of what might be created. You can anticipate steps, phases, surprises, synchronicities, challenges, and discoveries that move you toward completion and reflection. Generally, there is research and a gathering of materials and resources, which may involve interviewing colleagues or consulting your own notes from past projects. You begin by not just thinking about making but actually putting pen to page, fingers to computer keys. The beginning may feel good, like a bit of an accomplishment just to get over the inertia of preparing. A committed time schedule and place to work is essential. Even if you just write a few words or a paragraph each day, your mind and subconscious stay engaged. There is a thread of continuity that is unshakeable as you assemble the parts.

Somewhere midway, unless you are in one of those exceptional states where the work just flows, it becomes a muddle. What you thought was clear goes in too many directions. Colleagues' voices or your inner critic interrupt. There is an impulse to give up, to do something else with your time. And if you are practiced in the process or determined enough, you go back to your workspace and keep going. Resistance is part of the process of discovery. You need enough building blocks in place that the work can begin to tell you about itself. This is a time of following the energy of the piece. Insights come beyond what you could plan. Everything in your life has the possibility of impacting the outcome. There are revisions, reworkings, and completions, but know that this is just one phase.

Once you share the work with others—first readers or an editor—another phase emerges. What do you change and modify, and when do you stick to your original wording and dig deeper, even if it's unclear to others? Eventually there is the challenge of knowing where and when to end: How much is enough? Time deadlines can be both helpful and restrictive. Talking too much about your project can diminish or enhance your motivation. Everyone has a process that becomes familiar with repetition; growth and surprise are part of the aliveness of the work. And if you haven't learned something or been changed by your project, keep going even

You can't use up creativity. The more you use the more you have. — *Maya Angelou*

(*facing page*)
Whether (2013), sculpture by Rosalyn Driscoll
Wood, rawhide, 82 in. × 57 in. × 19 in.
Photograph by David Stansbury

Emptying
I was supposed to be in Singapore these weeks for a Fulbright Senior Scholar Program. When the residency was cancelled, there remained uninterrupted time. Something I've learned from my past: often you have to give something up to get something done. Creativity benefits from open space and time, even if it's not your desire.

if an external deadline has passed. It is worth the effort. This thing called "creativity" has a life of its own, from initial vision to completed form.

There is one certainty, you are not alone on the creative journey. Although you steer the boat, outer influences and collaborators are inherent to the process. How do you maintain a big view, credit others, and honor their participation while retaining your role as synthesizer and maker? There is also a process of letting go. Once your chapter is ready for print—art images found, words refined, "to do" explorations practiced—how do you make "empty" space so your next projects can emerge? These are interesting questions worth a lifetime of inquiry. As you complete your chapter, reflect on the ride. The end "product" may be close to your original vision, but the process will likely be a surprise.

Connections
Reciprocity is the practice of exchanging "things" with others for mutual benefit. As you shape your own chapter for this day, notice how everything around you feeds your imagination and interest. At various times, notice words, images, and experiences that might enhance a reader's interest and imagination. Notice how you find words for the experiences and facts you would like to communicate with others. In this process, you are both giving and receiving.

Interactive sculpture (2019)
Tate Modern, London, England

Mandy Kimm
Photograph by Alan Kimara Dixon

TO DO

TEN EMBODIMENT LANDMARKS
5 minutes

Seated, spine erect, with relaxed breathing:

- Take a moment and feel where your *feet* connect to the earth: open the portals of your feet, the soles of the feet.
- Stretch the palms of your *hands*. Open them to space.
- Yawn the back of your *throat*, releasing tension. These three gateways—feet, hands, throat—can open to the world or limit your connection.
- Find your *sitz bones*—the "feet" of your pelvis. Rock side to side to feel their location. Spread the diamond-shaped *pelvic floor* wide to help locate this region—the muscular-fascial floor of the pelvis.
- Locate the *sacrum* in back: the large bone linking the pelvic halves. Place your hands on the front and back of your belly and rock the sacrum forward and backward to feel its movement as part of the spine.
- Now, take your awareness downward to the very tip of your tail-bone, the *coccyx*, and give it a little wiggle. Imagine your favorite tail: dog, squirrel, dinosaur. Don't tuck it under, let it be free.
- Now, let's move your *spine* through lateral flexion. Say hello to your fish ancestors—move the tailbone (coccyx and sacrum) side to side and let the movement ripple up all your *spinal vertebrae*.

Noticing Ending

You never know when an ending will come. It is visceral. There is a shift in the air. It happens before you know it. The pen runs out of ink, the walk down the beach has a return, the window that was stuck is repaired. Your job is to notice . . .

- Greet your *tongue* and let it move. The tongue connects to the top of your spine through soft tissue. Let it wiggle and stretch, freeing any tension in this large, sensitive muscle that affects both speech and taste.
- To locate the *occipital condyles*—the feet of your skull, give a "yes nod" with your head. Touch your fingertips to the outside flaps of your ears and imagine a horizontal axis through the center of your skull. Rock the skull forward and backward, saying "yes", articulate just on the joint between skull and top vertebra, the cranial base.
- Stimulate the *center top of your skull* with your fingers. Staying grounded in your pelvis and feet, extend an imaginary string upward toward the sky, elongating the core of your body by connecting through ground and space.
- To finish your practice, make a diamond with your hands. (Thumbs and fingertips touching together with a diamond space between.) Place the thumbs in the center top of your head, and let this diamond shape connect you upward (like an antenna) to your favorite star—or the sun or moon.
- Spin the palms so they are parallel to the ground, and slowly descend your diamond hands downward in front of your body while you imagine an open spacious core. Relax and breathe deeply as you appreciate these ten landmarks of embodiment that help guide awareness inside the body, enhancing flow.

Refreshing Attention

On those days when I'd begin at dawn and still be at my computer writing at 2 p.m., our Brittany spaniel would pad upstairs to my desk and put both paws on the keyboard, insisting on a walk. Everyone should have a spaniel—a reminder to go outside and breathe. Sometimes it's hard to know when enough is enough. Some days it's clear.

TO WRITE

AUTHOR'S BIO

2 hours

Why speak about yourself? With so much "fake news" and hyperbole in the literary world, it is especially important for authors to reflect on and speak about their background. Begin by writing expansively. What concrete experiences underpin your writerly views and perspectives? What locations and cultural lenses impact the scope of your background—including home origins and landscapes, travel and work—leading to your present views? What multiple identities inform your lens; how were they formed, and when and why do you shift from one to the other? Concise, informative, and invitational writing about yourself settles a reader into the process of sharing the page.

After writing expansively, also create a short five-line author's bio that includes key points about your education, profession, location, and/or current focus.

Doug LeCours and Cameron McKinney
Middlebury, Vermont
*Body and Earth: Seven Web-Based
Somatic Excursions*
Still image from videography
by Scotty Hardwig

Acknowledgments

Honoring foundational colleagues: Nükhet Kardam for co-launching this project together, and somatic practices contributors Janet Adler, Bonnie Bainbridge Cohen, Lee Holden, and Caryn McHose.

Creative collaborators: Mary Abrams, Chris Aiken, April Danyluk, Scotty Hardwig, Shruthi Mahalingaiah, Rich Wolfson, and the many participating artists, scholars, scientists, and contributors named on the pages of this text.

Funders: Middlebury College and the Fund for Innovation, Faculty Development Fund, Digital Liberal Arts Grants, and Faculty Emeritus Fund; the Northampton Community Arts Council; and the Toad Fund for supporting events, travel, research, and films.

Wesleyan University Press and project editors: Suzanna Tamminen, director and editor-in-chief, for hosting this fourth book in my quartet of publications about the body; Hannah Crider and Jim Schley for detailed editing for the press; and Jane Crosen, Mandy Kimm, and Susan Prins for reviewing and editing throughout the multi-year process.

Previous publications: The editors and publishers who included and honed my articles and chapters in other publications leading toward this book: *Somatic Voices in Performance Research and Beyond*, edited by Christina Kapadocha for the Routledge Voice Studies series (2020); *Choreographic Practices: Performing Ecologies in a World in Crisis*, guest-edited by Sondra Fraleigh and Robert Bingham for Intellect, Inc. (2018); *Dancing on Earth*, edited by Kimerer LaMothe, with Sally Hess and Yvonne Daniel for the *Journal of Dance, Movement, & Spiritualities* (2017); *The Oxford Handbook of Dance and Wellbeing*, edited by Vicky Karkou, Sue Oliver, and Sophia Lycouris for Oxford University Press (2017); and *Sustainability*, edited by Jamie Devereaux for Mary Ann Liebert, Inc. (2013).

Chapter readers: Julia Basso, Harriet Brickman, Yvonne Daniel, Tangut Degfay, Rosalyn Driscoll, John Elder, Kimerer LaMothe, Joey Schmitz, Peter Schmitz, Laura Sewall, and Anne Woodhull.

Middlebury and Middlebury Institute colleagues: Dr. Beryl Levinger, Amy McGill, and Amy Collier, who supported many of the collaborative projects described in this text; and the Digital Learning Commons team Bob Cole, Mark Basse, and Evelyn Helminen, along with Daniel Houghton at Middlebury College who facilitated media documentation and creation.

Former students, workshop participants, and hosts: Each and every one of my fellow investigators and supporters over the last decade who have contributed to this book as well as creating your own offerings. This includes Body and Earth colleagues: Fabiano Culora, Eeva-Maria Mutka, and Susanna Recchia.

Friends and family: Especially my husband and co-creator, Stephen Keith.

THANK YOU ALL!

Heron Chest (1999) by Kristina Madsen
Pearwood, carved and gessoed
Photograph by David Stansbury

Notes

ENTRYWAYS: AN INTRODUCTION

Epigraph: Barry Lopez, "The Wild Road to the Far North," (Scott A. Margolin '99 Lecture in Environmental Affairs, Middlebury College, Apr. 12, 2007).

PREPARATIONS: USING THE TEXT

Epigraph: Janet Adler, "Where Are We Now?" (panel presentation, Somatics Festival 2019, Northampton, MA, filmed Sept. 21, 2019, 1:32:54), www.somatics2019.com/panel2.

DAY 1 BASIC CONCEPTS

Epigraph: E. B. White, "Intimations," in *One Man's Meat* (Thomaston, ME: Tilbury House Publishers, 2003), 220. In this essay, written in December 1941, just three days after the Pearl Harbor attacks, White writes: "Who is there big enough to love the whole planet? We must find such people for the next society."

1. The teepee is thought to be characteristic of the Gwich'in people, a group of Athabaskan-speaking North American First Nations people who inhabit the basins of the Yukon and Peel Rivers in eastern Alaska and Yukon—a land of coniferous forests interspersed with open, barren ground.

DAY 2 ATTITUDES

Epigraph: Bessel van der Kolk, "Trauma, Ecology, and Social Change" (panel presentation, The Embodiment Conference: Come Home to Your Body, Oct. 2020). See also Van der Kolk, *The Body Keeps the Score*.

1. Blakeslee and Blakeslee, *The Body Has a Mind of Its Own*, 11–12.

DAY 3 UNDERLYING PATTERNS: BIOLOGY

Epigraph: Emilie Conrad, "Letter from Emilie," Continuum Movement, accessed Dec. 9, 2021, continuummovement.com /letter-from-emillie/.

1. The distinction between biology and biography is common language among people who work with Silvan Tomkins's material. For an overview of principles, see *Affect Imagery Consciousness*. See also "Tomkins Institute, What Tomkins Said," accessed Dec. 22, 2021, www.tomkins.org/what-tomkins-said /introduction/affects-evolved-so-we-could-learn-what-to -seek-and-what-to-avoid.

2. For more on the three qualities of a cell, see Kapit and Elson, *The Anatomy Coloring Book*, 8.

3. Juhan, *Job's Body*, 21–34.

4. Bainbridge Cohen, *Sensing, Feeling, and Action*, 5.

5. For more on mirror neurons, see Vilayanur Ramachandran, "The Neurons That Shaped Civilization," Nov. 4, 2009 at TED-India, Myosore, India, video, 7:27, www.ted.com/talks /vilayanur_ramachandran_the_neurons_that_shaped _civilization?language=en.

6. "Open attention" is a phrase often used in Continuum Movement by founder Emilie Conrad. For an overview of Continuum, see Emilie Conrad, "Continuum: An Introduction," Jan. 24, 2013, YouTube video, 24:15, www.youtube.com /watch?v=IAacwbfveys.

7. To meet all the goats and artists/owners at Big Picture Farm, see "About Us," Big Picture Farm, accessed Dec. 8, 2021, bigpicturefarm.com/pages/about/.

DAY 4 UNDERLYING PATTERNS: BIOGRAPHY

Epigraph: Julia Alvarez, "Author Talk: Julia Alvarez" (MiddReads webinar, with Angela Evancie, Middlebury College, Middlebury, Vermont, Dec. 16, 2020).

1. To see this process demonstrated by Bonnie Bainbridge Cohen, see: Janet Adler, Bonnie Bainbridge Cohen, and Nancy Stark Smith, "Panel 1: Why Here, Why Then?" (panel presentation, Somatics Festival 2019, filmed Sept. 21, 2019 in Northampton, MA, 1:42:56), www.somatics2019.com/panel1.

2. Jon Hamilton, "Orphans' Lonely Beginnings Reveal How Parents Shape a Child's Brain," NPR, Feb. 24, 2014, www.npr .org/sections/health-shots/2014/02/20/280237833/orphans -lonely-beginnings-reveal-how-parents-shape-a-childs-brain. See also Juhan, "Skin: Touch as Food," in *Job's Body*, 43–44.

3. Ai Weiwei, *Circle of Animals/Zodiac Heads*, Yorkshire Sculpture Park, May 6, 2017–Nov. 1, 2020, visited June 2018.

4. Jung, *Psychological Types*. After emerging from a twenty-year process, Jung described this book as defining the ways in which his outlook differed from Freud's and Adler's.

5. For an overview of the work of neuroscientist Stephen Porges and his insights on the evolutionary origins of the social engagement system, differentiation of fight-or-flight, freeze- or-faint, and tend-and-befriend responses, and the skills to downregulate autonomic stress reactivity, see Porges, "The Emergence of the Social Engagement System: Insights from Evolution and Development," in *The Polyvagal Theory: Background and Criticism*, Polyvagal Institute, Aug. 2021, https:// www.polyvagalinstitute.org/background. For one of several additional models of response to stress and overwhelm that emphasize the "tend and befriend" response, see Bainbridge Cohen, *Sensing, Feeling, and Action*, 183. In "The Autonomic Nervous System: An Experiential Perspective," she writes: ". . . for women who get under this kind of stress, rather than *fff*, their response is to 'tend and befriend.'"

6. Brooks, *The Social Animal*, 148, 155; Lopez, *Horizon*, 42; Resmaa Menakem, "Notice the Rage; Notice the Silence," interview by Krista Tippett, *On Being*, June 4, 2020, audio, 50:53, onbeing.org/programs/resmaa-menakem-notice-the -rage-notice-the-silence/. Dan Siegel, "How our Relationships Shape Us," (lecture, Words on Wellness Speaker Series,

filmed Nov. 8, 2017, Jackson, Wyoming), YouTube video, 1:44:27, www.youtube.com/watch?v=fwmtgrWKQrY.

DAY 5 PERCEPTION

Epigraph: Bainbridge Cohen, *Sensing, Feeling, and Action*, 63–64.

1. Bainbridge Cohen, *Sensing, Feeling, and Action*.
2. Olsen, *Body and Earth*, 55–61.
3. Aaron Ebner and Adam Stieglitz discuss the Andean Alliance for Sustainable Development: Aaron Ebner and Adam Stieglitz, "Our Story," The Andean Alliance for Sustainable Development, accessed Aug. 21, 2021, alianzaandina.org/our-story/.
4. john a. powell and Stephen Menendian, "The Problem of Othering: Towards Inclusiveness and Belonging," *Othering & Belonging*, Issue 1 (Summer 2016): 14–39, www.otheringand belonging.org/the-problem-of-othering/.

DAY 6 ARRIVING AND ORIENTATION

Epigraph: Edward Albee, *The Zoo Story*, 1958 (first performed 1959).

1. Reflections on gravity and tonic function are from personal correspondence with Caryn McHose, following our Body and Earth workshop, Yorkshire Dales, Great Britain, 2017. Also see McHose and Frank, *How Life Moves*.
2. For more on Hubert Godard, see Kevin Frank, "Tonic Function: A Gravity Response Model for Rolfing Structural and Movement Integration," *Rolf Lines* (Mar. 1995): 12–20, alinenewton.com/pdf-articles/kevin-article.pdf.
3. Conversation with Professor Piri Sciascia, New Zealand Māori leader and Victoria University of Wellington (Te Herenga Waka) administrator. Research conducted during Fulbright Senior Scholar Fellowship to New Zealand at Whitireia Performing Arts, Feb.–May 2003.

DAY 7 ALIGNMENT

Epigraph: Tom Myers, "The Anatomy of Vitality: Self Practices for a Resilient Response to Our Challenges (presentation, The Embodiment Conference: Come Home to Your Body, Oct. 14–25, 2020). See also Thomas Myers, "Fascia 101," Nov. 20, 2014, YouTube video, 6:59, www.youtube.com/watch?v=-uzQMn87Hg0.

1. Houman and Flammia, *Intercultural Communication*, 160–89. Discussion on movement (kinetics), touch (haptics), space (proxemics), time (cronemics).
2. For foundational writings on posture and structure, see Ida Rolf, *Rolfing: The Integration of Human Structures* (New York: Harper and Row, 1978).
3. Olsen, *Bodystories*, 32. See also Doug Johnson and Melanie Sever, "Biotensegrity and Body Mapping" (presentation, 2017 Andover Educator's Conference, Association for Body Mapping Education, University of Redlands, Redlands, CA, revised June 2019), www.redlands.edu/globalassets/depts /music/body-mapping-conference-2019/johnson—sever —biotensegrity-and-body-mapping.pdf.

4. Bainbridge Cohen, *Sensing, Feeling, and Action*, 163. Also note that chiropractic treatment (and other spinal manipulation techniques), yoga, tai chi, and qigong are practices that encourage good spinal alignment.
5. For a model of social alignment, see R. Hari, T. Hindberg, L. Nummenmaa, M. Hamalainen, and L. Paarkkonen, "Synchrony of Brains and Bodies During Implicit Interpersonal Interaction," *Trends in Cognitive Sciences* 7, no. 3 (Feb. 4, 2013), doi.org/10.1016/j.tics.2013.01.003.

DAY 8 BREATH AND VOICE

Epigraph: Meklit Hadero, "The Unexpected Beauty of Everyday Sounds," Aug. 27, 2015, TED Fellows Retreat 2015, video, 12:54, www.ted.com/talks/meklit_hadero_the_unexpected_beauty _of_everyday_sounds.

1. For more on the spinning spirals of the human heart center, see Gil Hedley, "Unwinding the Heart Center," Jan. 15, 2021, YouTube video, www.youtube.com/watch?v=MbOyozg _GTs&feature=youtu.be.
2. Olsen, *Body and Earth*, 112–17.
3. References to the double dome of the diaphragm and the xiphoid process as the tail for the sternum are from: Bonnie Bainbridge Cohen, "Free the Vital Energy and Illuminating Presence of Your Spine: Explorations in Movement and Cellular Consciousness" (online workshop series, El Sobrante, CA, Oct. 1–Nov. 19, 2020).
4. See "Blood Basics," The American Society of Hematology, accessed June 14, 2021, www.hematology.org/education /patients/blood-basics: "Red blood cells begin as immature cells in the bone marrow, and after approximately seven days of development are released into the bloodstream."
5. For detailed information on vocal production, see Bonnie Bainbridge Cohen, *The Mechanics of Vocal Expression*, (El Sobrante, CA: Burchfield Rose Publishers, 2015), 13–22.

DAY 9 BALANCING THE NERVOUS SYSTEM

Epigraph: Bonnie Bainbridge Cohen, "Exploring the Embodiment of Cellular Consciousness through Movement: A Body-Mind Centering® Approach" (workshop, Facebook Live, June 11, 2020), www.facebook.com/watch/live/?v=268828757662712&ref =watch_permalink.

Note: This chapter includes contributions from neuroscientist Julia C. Basso, Feb. 1, 2021.

1. For more on the Moro reflex, see Bainbridge Cohen, *Sensing Feeling and Action*, 144.
2. Novelty is registered in general by the hippocampus and specifically the dentate gyrus. There are strong connections between the amygdala and hippocampus, which regulate emotional memory.
3. For details about the right and left brain, see Scaer, *8 Keys to Brain-Body Balance*, 22–25. The right brain is "intuitive, implicit, complex, affiliative, auditory, spatial, and creative" and the left brain is "verbal, mathematical, explicit, cognitive, linear, intellectual, and logical."

4. Weiwei Men, Dean Falk, Tao Sun, Weibo Chen, Jianqi Li, Dazhi Yin, Lili Zang, Mingxia Fan, "The Corpus Callosum of Albert Einstein's Brain: Another Clue to His High Intelligence?" *Brain* 137, no. 4 (Apr. 2014), https://doi.org/10.1093/brain/awt252.

5. See Porges, *The Polyvagal Theory* Also see Bainbridge Cohen, "Fight, Flight, Freeze, and Ease," (workshop, BMCA Conference on Resilience at Smith College, June 2018), www.bodymindcentering.com/. "Fight, flight, and freeze, as patterns of movement, share a common root of dis-ease that gives rise to different spatial responses." In fight, movement is directed spatially in the direction of a perceived oncoming force.

6. See more on gut-brain interconnectivity in Mayer, *The Mind-Gut Connection*.

7. See Dr. Dan Siegel, "Dr. Dan Siegel's Hand Model of the Brain," Aug. 9, 2017, YouTube video, 8:19, www.youtube.com/watch?v=f-m2YcdMdFw. See also Scaer, *8 Keys to Brain-Body Balance*, 11: "If you make a model of your brain by folding your thumb against your palm and covering it with your curled fingers, the fingers reflect the cortex, the thumb becomes the limbic system, and your wrist signifies the brainstem."

8. For new research on the cerebellum and cognitive-emotional processes, see Jeremy Schmahmann, "The Cerebellum and Cognition," *Neuroscience Letters*, 688 (2019): 62–75.

DAY 10 DANCING WITH THE ENDOCRINE SYSTEM

Epigraph and writing collaboration from correspondence with Dr. Shruthi Mahalingaiah, June 22 and July 14, 2020. For biographical information, see: "Shruthi Mahalingaiah," Harvard School of Public Health, www.hsph.harvard.edu/shruthi-mahalingaiah/.

1. See National Women's Health Network Website for specifics on toxins in cosmetics and personal care products: M. Isabelle Chaudry, "Toxic Personal Care Products and Women's Health: A Public Health Crisis," National Women's Health Network, Aug. 5, 2019, accessed Aug. 4, 2020, www.nwhn.org/toxic-personal-care/.

2. The heart secretes a variety of hormones and acts in both a paracrine and endocrine way including the secretory phenotype of mammalian atrial cardiomyocytes that is associated with the production of the natriuretic polypeptide hormones ANF and BNP.

3. Bainbridge Cohen, *Sensing, Feeling, and Action*, 54–62.

4. See Patty Townsend, "Radiance and Levity—The Glandular System in Yoga," Embodyoga Blog, Oct. 5, 2017, www.embodyogablog.com/2017/10/05/the-glandular-system-in-yoga/.

5. Bainbridge Cohen, *Sensing, Feeling, and Action*, 57.

DAY 11 MOVEMENT AND TOUCH

Epigraph: Anna Halprin with Rachel Kaplan, *Making Dances That Matter: Resources for Community Creativity* (Middletown, CT: Wesleyan University Press, 2019), 12.

1. See Olsen, "Being Seen, Being Moved" in *The Place of Dance*, 42–43 and 187–89.

2. For touch research, see Juhan, *Job's Body*, 43–57.

3. See *Contact Quarterly*: "About Contact Improvisation (CI)—Definitions," accessed Jan. 10, 2020, contactquarterly.com/contact-improvisation/about. "Contact Improvisation is an evolving system of movement initiated in 1972 by American choreographer Steve Paxton. The improvised dance form is based on the communication between two moving bodies that are in physical contact and their combined relationship to the physical laws that govern their motion—gravity, momentum, inertia."

4. Bainbridge Cohen, *Sensing, Feeling, and Action*, 1. And: Bainbridge Cohen, *Basic Neurocellular Patterns: Exploring Developmental Movement*, 129–31.

DAY 12 SPEAKING AND LISTENING

Epigraph: Hubert Godard, "Reading the Body in Dance: A Model," *Rolf Lines/Journal of the Rolf Institute* 22 (1994): 36–41.

1. For more specifics on autonomic nervous system responses, see Porges, *The Polyvagal Theory*.

DAY 13 WRITING AND READING ALOUD

Epigraph: Mary Karr, "Sacred Carnality," *New Yorker*, Oct. 11, 2015, 71–78, www.newyorker.com/books/page-turner/sacred-carnality.

1. Karr, *The Art of Memoir*, 72.

2. Barry Lopez, "The Wild Road to the Far North" (writing workshop following lecture, Middlebury College, Apr. 13, 2007). Also listen to: "The Archive Project, Barry Lopez: *Horizon*," Literary Arts Archive, Mar. 19, 2019, podcast, 1:03:24, literary-arts.org/archive/barry-lopez-horizon/.

3. Joan Borysenko (trained as both a cell biologist and clinical psychologist) from: "Writing Down the Light" (workshop, Kripalu Center for Yoga & Health, Stockbridge, MA, Nov. 16–18, 2018).

4. For an invitation into the writing process, see Ken Macrorie, *Telling Writing*.

5. Natalie Goldberg, *Writing Down the Bones: Freeing the Writer Within* (Boulder, Colorado: Shambala Publications, 2005).

6. Personal correspondence with John Elder, Professor emeritus of English and American Literatures and Environmental Studies at Middlebury College, Vermont, Mar. 8, 2020.

7. From E. B. White, Edmund Ware Smith, and Martha White, *Chickens, Gin, and a Maine Friendship: The Correspondence of E. B. White and Edmund Ware Smith* (Guildford, CT: Down East Books, 2020), 134.

8. "Bill T. Jones, Action/Phase 1: Reading," New York Live Arts, accessed June 13, 2020, newyorklivearts.org/event/action-phase-1-reading/?.

9. Eve Sedgwick, *Touching Feeling: Affect, Pedagogy, Performativity* (Durham, NC: Duke University Press, 2003). Also John Updike, "Novelist John Updike," interview by Terry Gross, *Fresh Air*, NPR, Oct. 14, 1997, audio, freshairchive.org/segments/novelist-john-updike. Samuel Johnson

(1709–1784) is featured in the Quote Fancy website, accessed Dec. 10, 2021, quotefancy.com/quote/988262/Samuel-Johnson-If -you-want-to-be-a-writer-then-write-Write-every -day. The quote is drawn from Samuel Johnson, *The Major Works*, ed. Donald J Greene (London: Oxford University Press, 2000).

DAY 14 EMBODYING DIGITAL MEDIA AND FILMING

Epigraph: Gardner Campbell, "Romantic Computing" (lecture, Middlebury Institute of International Studies, May 2017).

Writing collaboration with Scotty Hardwig is from correspondence, May–Aug., 2020. For biographical information and current media works, see Anatomy/Zero, "About," anatomyzero .com/about.

1. See Amy Collier, "Embodiment and the Digital: Constructing and Deconstructing the Body" (Embodied Intelligence Lecture and Workshop Series, Middlebury Institute of International Studies, Monterey, CA, Feb. 17, 2017), sites.middle bury.edu/embody/.
2. Blakeslee and Blakeslee, *The Body Has a Mind of Its Own*, 130.
3. Collier, "Embodiment and the Digital."
4. The history of these media biases can be traced back to the turn of the twentieth century during the height of industrialism in the West. In his documentary series, *The Century of the Self*, filmmaker Adam Curtis traces the career of PR consultant Edward Bernays, who was one of the godfathers of contemporary public relations and mass media advertising. Using Freud's theories of need, desire, and sexuality, Bernays harnessed the power of imagery to sell a variety of products, from cars to cigarettes. Paul Mazur, a leading Wall Street banker at the time, is cited as saying that "we must shift America from a needs to a desires culture. People must be trained to want new things, even before the old have been entirely consumed. Man's desires must overshadow his needs." This philosophical shift marked a new era in human communication, consumption, and mass imagery.
5. Richard Lanham, *The Economics of Attention: Style and Substance in the Age of Information* (Chicago: University of Chicago Press, 2007).
6. See Delaney and Schroeppel, *The Bare Bones Camera Course for Film and Video* for foundational information on filming.
7. Gardner Campbell, "Romantic Computing" (lecture, Middlebury Institute of International Studies, Monterey, CA, May 2017).
8. See Scotty Hardwig, "Five Central Elements for Video Makers," filmed in Carmel and Monterey, CA, 2018, 7:23, andrea -olsen.com/films/.

DAY 15 ART MAKING AND IMAGINING

Epigraph: Gordon Thorne, Personal Notebook. For biographical information and artworks by Gordon Thorne, see APE, "About," www.apearts.org/gordon-thorne.html.

1. "Bill T. Jones: A Life Well-Danced," interview by Onye Ozuzu, Chicago Humanities Festival, Oct. 24, 2013, YouTube video, 1:03:07, www.youtube.com/watch?v=QfvUYqJjQbs.

2. For description of the Te Māori exhibit at the Metropolitan Museum of Art, see "Te Māori exhibition opens in New York," New Zealand History, Sept. 10, 1984, nzhistory.govt.nz/te -maori-exhibition-opens-in-new-york.
3. Yo-Yo Ma quote from "The Weaponization of Culture," (panel presentation, World Economic Forum/Davos, Davos-Klosters, Switzerland, Jan. 23, 2018), www.youtube.com /watch?v=oNCEPHCmgZA.
4. Philip Buller, "Bodies of Work—Beach, 2014–15," accessed June 11, 2020, www.philipbuller.com.
5. For information on the NES Artist Residency, see "About NES," accessed Dec. 20, 2021, neslist.is/about-nes/.
6. See Buller, www.philipbuller.com.
7. Gordon Thorne, "Vision of Thornes," interview by Scotty Hardwig, 2017, APE, video, www.apearts.org/gordon-thorne .html. See also Richard Florida, *The Rise of the Creative Class*, 2nd ed. (New York: Basic Books, 2012).
8. Paul Taylor, *Dancemaker*, directed by Matthew Diamond (New York: Artistic License Films, 1998) documentary film, www.youtube.com/watch?v=gj4q_LhYICk.
9. Spoken by the playwright Tony Kushner in *Wrestling with Angels*, Freida Lee Mock's documentary portrait reviewed in the *New York Times* by Stephen Holden, Oct. 4, 2006, www .nytimes.com/2006/10/04/movies/04ange.html.

DAY 16 FACE AND EXPRESSION

Epigraph: Chimamanda Ngozi Adichie, *Purple Hibiscus*, 226.
1. See Dr. Minass, "Embryology of the Face (Easy to Understand)," Apr. 21, 2019, YouTube video, www.youtube.com /watch?v=cVPMjE2UA70.
2. The facial bones include the vomer, two nasal conchae, two nasal bones, two maxilla, the mandible, two palatine bones, two zygomatic bones, and two lacrimal bones; seven of these comprise each eye socket.
3. Sources vary about the number of muscles in the face. Most say there are forty-three "facial action units," others identify eighty, and the TV show *Lie to Me* claims ninety-one.
4. See Judy Foreman, "A Conversation with Paul Ekman: The 43 Facial Muscles That Reveal Even the Most Fleeting Emotions," *New York Times* (online), Aug. 5, 2003, www.facebook.com /watch/?v=10201589807416055. For more about micro-expressions, see "Microexpressions," Paul Ekman Group, accessed Dec. 10, 2021, www.paulekman.com/resources/micro -expressions/.
5. The words multilingual and linguistic derive from the Latin root word lingua, meaning "tongue."
6. Mevlâna Jalâluddîn Rumi, *The Life and Spiritual Milieu of Mev Mevlâna Jalâluddîn Rumi*, trans. Camille Helminski and Kabir Helminski. (Louisville, KY: Threshold Books, 2017).

DAY 17 ENERGY AND VITALITY

Epigraph: W. H. Hudson (British author, naturalist, and ornithologist, 1841–1922), *The Book of a Naturalist* (London: J. M. Dent, 1924).

Writing collaboration with Rich Wolfson from correspondence Apr. 17, 2020. For biographical information, see: "The Great Courses, Richard Wolfson, Ph.D.," www.thegreatcourses.com/professors/richard-wolfson/.

1. Einstein's $E = mc^2$ can be viewed as showing that energy has mass.

2. For more on waves and water, see Richard Wolfson, "Let There Be Light" in *Simply Einstein: Relativity Demystified* (New York: W. W. Norton & Company, 2003), 36–54. Note that water molecules themselves are in constant motion due to their thermal energy.

3. In physics, "power" is not interchangeable with "energy." "Power" is the rate of energy use, production, generation, or loss.

4. For more on *wei qi* and the immune system, see Lee Holden, "Immunity Emergency Kit," Holden Qigong, accessed Mar. 15, 2021, www.holdenqigong.com/15-minute-immune-system-booster/.

5. Attributed to novelist Raymond Chandler, "The Simple Art of Murder," in *Saturday Review of Literature* (New York: Saturday Review Associates), 13–14.

DAY 18 HUMOR AND LAUGHTER

Epigraph: E. B. White and Katharine S. White, eds., *A Subtreasury of American Humor* (New York: Coward-McCann, 1941), xvii.

Writing collaboration from correspondence with April Danyluk, Apr. 14, 2020; drawn from her essay "Employing Humor for Social Change: A Theoretical and Practical Exploration" (unpublished manuscript, Monterey Institute of International Studies at Monterey, CA, 2016). For biographical information, see: "The World According to Student April Danyluk," Apr. 2, 2016, www.middlebury.edu/institute/news/world-according-student-april-danyluk.

1. For one of several videos showing babies and laughter response, see mandkyeo, "Emerson—Mommy's Nose is Scary!" YouTube video, 0:58, www.youtube.com/watch?v=N9oxmRT2YWw.

2. See Dr. Lee Berk, "Gelatology/Laughter," interviewed by Alie Ward, *Ologies*, Feb. 5, 2018, audio, www.alieward.com/ologies/gelotology.

3. See Dr. Fry, "The Science of Laughter," Nov. 10, 2007, video, www.exploratorium.edu/video/science-laughter.

4. See Ramon Mora-Ripoli, "The Therapeutic Value of Laughter in Medicine," *Alternative Therapies in Health and Medicine* 16, no. 6 (2010): 56–64. See also Laughter Online University, "Ramon Ripoll," accessed June 10, 2020, www.laughteronlineuniversity.com/rmora-ripoll/.

5. Anne Tyler, *Clock Dance* (New York: Knopf, 2018), 8.

6. Scripture quoted from the New King James version of the Bible, Proverbs 17:22 (English translation). For international research on rhythmic laughter, see Radoslaw Niewiadomski, Maurizio Mancini, Yu Ding, Catherine Pelachaud, and Gualtiero Volupa, "Rhythmic Body Movements of Laughter,"

Conference Paper, International Conference on Multimodal Interaction, Nov. 2014, www.researchgate.net/publication/268925736_Rhythmic_Body_Movements_of_Laughter.

7. See Berk interview.

8. Mahadev Apte, *Humor and Laughter: An Anthropological Approach* (Ithaca, NY: Cornell University Press), 1985.

9. *Philogelos* (*The Laughter Lover*) is a collection of 265 jokes, likely made in the fourth or fifth century CE (no longer in print).

DAY 19 RESISTANCE AND RELATIONSHIP

Epigraph: Barbara Dilley, quoted in: Rachel Avev, "The Unravelling of a Dancer," *New Yorker*, Mar. 30, 2020.

Interview and writing collaboration with Chris Aiken, Northampton, MA, July 2020. For biographical information see: "Chris Aiken," Smith College, www.smith.edu/academics/faculty/chris-aiken.

1. David Bohm, *On Dialogue*, 2nd ed. (New York: Routledge, 2004).

DAY 20 SENSATIONS AND EMOTIONS

Epigraph and writing collaboration: Mary Abrams on the affective system, communication, and somatics, interviewed by Andrea Olsen in Holderness, NH, July 2019. For biographical information, see: "About Mary Abrams," Moving Body Resources, movingbodyresources.com/about-mbr/about-mary-abrams/.

1. Tomkins, *Affect, Image, Consciousness*. Tomkins identified nine innate affects that humans possess and from these, developed a set of four highly specific behavioral requirements known as "The Tomkins Blueprint for Individual Mental Health." See also "What Tomkins Said," The Tomkins Institute, accessed Dec. 18, 2021, www.tomkins.org/what-tomkins-said/.

2. Milton Bennett and Ida Castiglioni, "Embodied Ethnocentricism and the Feeling of Culture: A Key to Training for Intercultural Competence," IDRInstitute, 2004, www.idrinstitute.org/resources/embodied-ethnocentrism-feeling-culture/.

3. For more information about the term "silent-level processes," related concepts, and general semantics based on the work of A. Korzybski (1879–1950), see George Doris, "Korzybski and General Semantics," *ETC: A Review of General Semantics* 62, no. 4 (Oct. 2005): 421–423. See also J. S. Bois, *The Art of Awareness: A Textbook on General Semantics*, (Dubuque, IA: W. C. Brown, 1966). Bois worked closely with Korzybski.

4. For more on emotion, see Damasio, *The Feeling of What Happens*.

5. Doidge, *The Brain's Way of Healing*.

6. Levine, "The Art of Noticing." For more on Somatic Experiencing and resources on healing from trauma, see: Somatic Experiencing International, "Research and Resources," accessed Dec. 10, 2021, traumahealing.org/se-research-and-articles/.

DAY 21 VISION AND INSIGHT

Epigraph: Maxine Green, "Toward Wide-Awakeness: An Argument for the Arts and Humanities in Education." Published in "Issues in Focus: The Humanities and the Curriculum," *Teacher's College Record* 79, no. 1 (Sept. 1977): 119–25.

1. To meet Ida Bagus Anom Suryawan and learn about Barong masks and their connections to trees, see Culture Trip, "The Sacred Meanings of Balinese Masks" YouTube video, www.youtube.com/watch?v=WrczC8_JO2c.

2. For information on the embryological origins of the eyes and their connections to the brain and stress, see Jessica Wapner with neurobiologist Andrew Huberman, "Vision and Breathing May be the Secrets to Surviving 2020," *Scientific American*, Nov. 16, 2020, www.scientificamerican.com/article/vision-and-breathing-may-be-the-secrets-to-surviving-2020.

3. See V. S. Ramachandran, "Filling in Gaps in Perception: Part I, Current Directions in Psychological Science," *Sage Publications* 1, no. 6 (Dec. 1992): 199–205.

DAY 22 IDENTITY AND FLUIDITY

Epigraph: Daniel J. Siegel, *Mindsight*.
Writing collaboration from conversations and book retreat with Nükhet Kardam while co-teaching at the Middlebury Institute of International Studies at Monterey, CA, 2016–2018. For biographical information, see: Middlebury Institute of International Studies, "Nükhet Kardam," www.middlebury.edu/institute/people/nukhet-kardam.

1. Performances at the Yerba Buena Center for the Arts in San Francisco, 2013, 2018.

2. See Katie Shonk, "What is Conflict Resolution and How Does it Work," Program on Negotiation, Harvard Law School Daily Blog, Oct. 19, 2020, accessed Feb. 14, 2021, blog post, www.pon.harvard.edu/daily/conflict-resolution/what-is-conflict-resolution-and-how-does-it-work/.

3. Kevin Giordano, "Going to the Wall," *Dance Magazine*, May 1, 1999. See also: Bebe Miller Company, "Works by Bebe Miller and Company," accessed June 10, 2021, bebemilleruwmhistory.weebly.com/works-of-art.html

4. Conversation with Fabiano Culora, Lecturer in Performance and Bodywork Supervisor at Northern School of Contemporary Dance, Leeds, Great Britain, Jan. 25, 2021. He adds, "We can rehearse/host a role through art practice, yielding more choice to stay present in our wider life-context communications."

DAY 23 SPIRITUALITY AND EMBODIMENT

Epigraph: Reginald Ray, *Touching Enlightenment*, 12. He continues: "In this, I am speaking not of the body we think we have. Rather, I am talking about the body that we meet when we are willing to descend into it, to surrender into its darkness and its mysteries, and to explore it with our awareness."

1. Thomas Berry, *The Dream of the Earth* (Berkeley, CA: Counterpoint Press, 2015).

2. "There are wide variations among people as to just what reality contains." For this quote and discussion of the complexities involved in defining religion and religions, see Rodrigues and Harding, *Introduction to the Study of Religion*, 1–16. Also see Craig Martin, *A Critical Introduction to the Study of Religion* (New York: Taylor & Francis, 2017).

3. "Mystical experience was defined by Harvard professor Dr. William James over a hundred years ago. He identified key aspects: a brief interlude beyond time and space that feels realer than real, ineffable—or impossible to put into words; imparting precious wisdom unavailable by other means; and a gift of grace rather than a result of will." Also see, from Carolyn Rivers, "Creating Circles of Support" (webinar with Joan Borysenko, The Sophia Institute, Charleston, South Carolina, Apr. 18, 2020), thesophiainstitute.org/events/joan-borysenko-may-1-2020/.

4. See Janet Adler, "Where Are We Now?" (panel presentation, Somatics Festival 2019, Northampton, MA, 2019), www.somatics2019.com/panel2.

5. There are over 170 translations of the *Tao te Ching* (or *Book of the Way*) by Lao Tzu; the translation referenced here is by Stephen Mitchell, trans., *Tao te Ching: A New English Version* (New York: Harper Collins, 1991).

6. Shunryu Suzuki, *Zen Mind, Beginner's Mind: Informal Talks on Zen Meditation and Practice* (Berkeley, CA: Shambala, 2011).

7. For more on dimensions of contemplative practice, linking daily life and spiritual practice, see Dilley, *This Very Moment*.

8. From *Armenia!* exhibit, 2018, Metropolitan Museum of Art, New York. Richly illuminated manuscripts created at the scriptorium at Haghpat Monastery in Armenia show many images of angels whispering in the ears of the writers.

9. For more on enchantment, see Abram, *The Spell of the Sensuous*.

10. Personal conversation with Bonnie Bainbridge Cohen, Aug. 15, 2019.

11. See Janet Adler, "Where Are We Now?" (panel presentation, Somatics Festival 2019, Northampton, MA, 2019), www.somatics2019.com/panel2.

DAY 24 REST AND RESTORATION

Epigraph: Pema Chödrön, "Three Methods for Working with Chaos." *Lion's Roar* magazine, Sept. 17, 2021, www.lionsroar.com/pema-chodrons-three-methods-for-working-with-chaos/. "Go to the places that scare you, use poison as medicine, and regard what arises as awakened energy."

1. Dr. Herbert Benson with Miriam Klipper, *The Relaxation Response* (New York: Avon/Harper Collins, 1976).

2. See John Peever Murray and Brian J. Murray (directors of sleep laboratories), "What Happens in the Brain During Sleep?" *Scientific American*, Sept. 2015, www.scientificamerican.com/article/what-happens-in-the-brain-during-sleep1/. Current research reports that two distinct types of sleep are generated by the brain: slow-wave sleep (SWS), known as deep sleep, and rapid eye movement (REM), also

Jonah Keith
Pfeiffer Beach,
Big Sur, California
(2017)
Photograph by
Stephen Keith

called dreaming sleep. Most of the sleeping we do is of the slow-wave variety (also called non-REM sleep).

3. See Matthew Walker, *Why We Sleep: Unlocking the Power of Sleep and Dreams* (New York: Simon & Schuster, 2017). Also see book website, accessed Dec. 10, 2021, www.simonand schuster.com/books/Why-We-Sleep/Matthew-Walker /9781501144325. "Within the brain, sleep enriches our ability to learn, memorize, and make logical decisions. It recalibrates our emotions, restocks our immune system, fine-tunes our metabolism, and regulates our appetite. Dreaming mollifies painful memories and creates a virtual reality space in which the brain melds past and present knowledge to inspire creativity."

4. Jung, *Psychological Types*. Also see Jung's writing on the Anima and Animus within each person in volume 9 of his collected works: "The Syzygy: Anima and Animus," Vol. 9, Part 2.

5. Levine, *In an Unspoken Voice*.

6. Constructive rest is discussed by Mabel Todd in her classic book *The Thinking Body: A Study of the Balancing Forces of Dynamic Man*, originally published in 1937. Reprint edition: Highland, NY: Gestalt Journal Press, 2015.

DAY 25 FOOD AND RITUAL

Epigraph: M. F. K. Fisher, *The Art of Eating*, 50th Anniversary Edition (New York: Houghton Mifflin Harcourt, 2004).
Writing conversation with Harriet Brickman, Jan. 12, 2021.

1. For discussion of ritual as a repeated series of action-centered practices, see Hillary Rodrigues and John Harding, *Introduction to the Study of Religion*, 16: "Ritual is a complex category of human behavior, not found exclusively within what is clearly accepted as religious activity."

2. See discussions of food and contemporary life. Michael Pollan, *In Defense of Food*.

3. See Amy Qin, "Coronavirus Threatens China's Devotion to Chopsticks and Sharing Food," *New York Times* (online), May 25, 2020, www.nytimes.com/2020/05/25/world/asia /china-coronavirus-chopsticks.html?smid=nytcore-ios-share.

4. Mayer, *The Mind-Gut Connection*. See also Sandra Blakeslee, "Complex and Hidden Brain in the Gut Makes Cramps, Butterflies, and Valium," *New York Times*, Jan. 23, 1996. This was the first essay that stimulated my imagination into the multilayered impacts of the digestive system.

5. From the Territorial Seed Company Catalog, Cottage Grove, OR, 2020.

DAY 26 MIND AND MINDFULNESS

Epigraph: Saint Augustine, *The Confessions: Saint Augustine of Hippo*, ed. David Vincent Meconi (San Francisco, CA: Ignatius Press, 2012).

1. For an overview of research on brain-heart-gut interconnectedness, see Robert Scaer, *8 Keys to Brain-Body Balance*.

2. Olsen, *Body and Earth*, 39–48.

3. See Kenneth Cohen, *The Way of Qigong*, 319–31. "Prenatal qi is the constitutional essence with which we are born, and postnatal qi is our basic daily energy that we cultivate through our diet and lifestyle."

4. Mayer, *The Mind-Gut Connection*.

5. For more on effective posture in contemplative practice, see Erika Berland, *Sitting: The Physical Art of Meditation* (Boulder, CO: Somatic Performer, 2017).

6. For twenty-nine years of research, articles, and training programs about the intelligent heart, see the Heart Math Institute's Research Library, accessed Dec. 10, 2021, /www .heartmath.org/research/research-library/.

7. Peter Schjeldahl, "The Art of Dying," *New Yorker*, Dec. 16, 2019, www.newyorker.com/magazine/2019/12/23/the-art -of-dying.

8. For more on the commodification of mindfulness and the silencing of Asian and Asian American Buddhists in the teaching of mindfulness see Funie Hsu, "What Is the Sound of One Invisible Hand Clapping? Neoliberalism, the Invisibility of Asian and Asian American Buddhists, and Secular Mindfulness in Education" in *Handbook of Mindfulness: Culture, Context, and Social Engagement*, eds. Ronald E. Purser, David Forbes, and Adam Burke (New York: Springer Publishing, 2016), 369–81. Also see the writings of Edwin Ng and the "engaged Buddhism" movement. Other resources include David McMahan and Erik Braun, "From Colonialism to Brainscans: Modern Transformations of Buddhist Meditation" in *Meditation, Buddhism, and Science* (New York: Oxford University Press, 2017), 1–15.

9. From *Awakening Grace: Six Somatic Tools*, text and performance by Andrea Olsen, Smith College, Northampton, MA, Dec. 2019.

DAY 27 BODY-MIND CENTERING®

Epigraph: Bonnie Bainbridge Cohen, *Sensing, Feeling, and Action*.

1. This introduction and the related quotes are used with permission from Bonnie Bainbridge Cohen. Body-Mind Centering® is a registered service mark, and BMCᴿᴹ is a

service mark of Bonnie Bainbridge Cohen. For information about the Body-Mind Centering® approach and The School for Body-Mind Centering's programs and courses, visit https://www.bodymindcentering.com/. For information about Bonnie Bainbridge Cohen, her books, videos, and events, visit https://bonniebainbridgecohen.com/.

2. Note from Bonnie: "I use the word 'somatization' to engage the kinesthetic experience directly, in contrast to 'visualization,' which utilizes visual imagery to evoke a kinesthetic experience. Through somatization, the body cells inform the brain as well as the brain informs the cells. I derived the word 'somatization' from Thomas Hanna's use of the word 'soma' to designate the experienced body, in contrast to the objectified body. When the body is experienced from within, the body and mind are not separated but are experienced as a whole. While Tom spoke of this during the 1960s, his first book utilizing the term 'soma' was *Bodies in Revolt*, which came out in 1970. Tom coined the term 'somatics' in 1976 when he founded and named the *Somatics Magazine: Journal of the Bodily Arts and Sciences.* Somatics also names a field of study—the study of the body through a personal, experiential perspective. Body-Mind Centering is a small part of this burgeoning field."

3. Numbers in this paragraph were updated in phone conversation with Bainbridge Cohen, Jan. 30 and Dec. 20, 2021.

4. Bonnie Bainbridge Cohen, "Exploring the Embodiment of Cellular Consciousness Through Movement: A Body-Mind Centering Approach" (online course, Summer Series 2020, Aug. 6, 2020). As of January 2021, the number of countries represented in the online course was ninety.

5. Bainbridge Cohen also notes that "the nearby gallbladder, which stores bile for digestion, gives a sense of lightness in movement."

6. For information about the embryological origins of the gonads and the descent of the testes through the deep body wall to the scrotum, see Frank H. Netter, *Atlas of Human Anatomy*, 7th ed. (Amsterdam, Netherlands: Elsevier Press, 2018), 179. Also see Lauren J. Sweeney, *Basic Concepts in Embryology: A Student's Survival Guide*, 1st ed. (New York: McGraw-Hill, 1997).

DAY 28 THE DISCIPLINE OF AUTHENTIC MOVEMENT

Epigraph: Janet Adler, "Where Are We Now?" (panel presentation, Somatics Festival 2019, Northampton, MA 2019), www.somatics2019.com/panel2.

1. This brief description and related quotes are used with permission from Janet Adler and the Circles of Four faculty, Dec. 16, 2021. For detailed information, see: "Circles of Four—Discipline of Authentic Movement," disciplineofauthentic movement.com/.

2. Janet Adler, "The Mandorla and the Discipline of Authentic Movement," *Journal of Dance & Somatic Practices* 7, no. 2 (2015): 217–27.

3. See Janet Adler, "Intimacy and Emptiness: The Evolution of the Inner Witness in the Discipline of Authentic Movement," (workshop, Somatics Festival 2019, Northampton, MA, Sept. 21, 2019), www.somatics2019.com/authenticmovement.

DAY 29 QIGONG

Epigraph and writing collaboration from Lee Holden 2010 and correspondence Mar. 6, 2020. For biographical information, see: Holden Qigong, "How Lee Learned Qigong," accessed Dec. 10, 2021, www.holdenqigong.com/how-lee-learned-qi-gong/.

1. Holden, *Qigong for Health and Healing*, 10.

2. For more on the lineage of qigong, see Holden Qigong, "History of Qigong," www.holdenqigong.com/history-of-qigong/.

3. For historical background on the variations of spelling qigong, its translations, and on the yin/yang symbol, see Kenneth Cohen, *The Way of Qigong*, 7, 30–31.

4. Personal conversation with Amy Kiara Ruth about yin/yang symbol and complementarity, Feb. 4, 2021.

5. To experience Chungliang Al Huang in action, see his TEDx talk, "Tai Ji/YinYang Philosophy," TEDxHendrixCollege, Conway, AR, May 22, 2012, YouTube video, 30:50 www.youtube.com/watch?v=8TSEn0Aa39s.

DAY 30 BODY AND EARTH

Epigraph: David Abram, "Between the Human Imagination and the Land's Wild Dreaming"(presentation, The Embodiment Conference: Come Home to Your Body, Oct. 14–25, 2020).

Writing collaboration with Caryn McHose is from three decades of co-teaching. For biographical information, see: "Resources in Movement—About the Co-Founders," resources inmovement.com/about-the-co-founders/.

1. Woven into the fabric of this practice are experiences drawn from work with Emilie Conrad and Susan Harper, founders of Continuum; Bonnie Bainbridge Cohen of the School for Body Mind Centering®; the discipline of Authentic Movement as developed by Janet Adler; Peter Levine's work in Somatic Experiencing®; Ida Rolf and the Rolf Institute in Rolfing® and Rolf Movement®; and dance improvisation and performance in its many dimensions.

2. Caryn McHose, interviewed by Donnella Wood, Body and Earth Retreat, Holderness, NH, July 19, 2019.

DAY 31 YOUR CHAPTER

Epigraph: Maya Angelou, "Profile" in *Bell Telephone Magazine*, 61, no. 1, 1982.

SELECTED BIBLIOGRAPHY AND WEB RESOURCES

Epigraph: Saint Augustine, *The Confessions: Saint Augustine of Hippo*, ed. David Vincent Meconi (San Francisco, CA: Ignatius Press, 2012).

Selected Bibliography and Video Resources

Our use of words is generally inaccurate and seldom completely correct, but our meaning is recognized none the less. — Saint Augustine, *from* The Confessions

Shimmer / Symudliw (2020) by Eeva-Maria Mutka
Pen Pynfarch, Wales
Photograph from installation

Abram, David. *The Spell of the Sensuous: Perception and Language in a More Than Human World*. New York: Vintage, 1997.

Adler, Janet. "The Mandorla and the Discipline of Authentic Movement." *Journal of Dance & Somatic Practices* 7, no. 2 (2015): 217–227.

Adler, Janet. *Offering from the Conscious Body: The Discipline of Authentic Movement*. Rochester, VT: Inner Traditions, 2002.

Albright, Ann Cooper. *How to Land: Finding Ground in an Unstable World*. New York: Oxford University Press, 2018.

Alvarez, Julia. *Afterlife*. Chapel Hill, NC: Algonquin Books, 2020.

———. *Something to Declare*. Chapel Hill, NC: Algonquin Books, 1998.

Bainbridge Cohen, Bonnie. *Basic Neurocellular Patterns: Exploring Developmental Movement*. El Sobrante, CA: Burchfield Rose Publishers, 2018.

———. *Sensing, Feeling, and Action: The Experiential Anatomy of Body-Mind Centering*, 3rd ed. Northampton, MA: Contact Editions, 2012.

Blakeslee, Sandra, and Matthew Blakeslee. *The Body Has a Mind of Its Own: How the Body Maps in Your Brain Help You Do (Almost) Everything Better*. New York: Random House, 2007.

Brooks, David. *The Social Animal: The Hidden Sources of Love, Character, and Achievement*. New York: Random House, 2012.

Cohen, Kenneth S. *The Way of Qigong: The Art and Science of Chinese Energy Healing*. Canada: Wellspring/Ballantine Books, 1999.

Conrad, Emilie. *Life on Land: The Story of Continuum*. Berkeley, CA: North Atlantic Books, 2007.

Damasio, Antonio. *The Feeling of What Happens: Body and Emotion in the Making of Consciousness*. New York: Harcourt Brace, 1999.

Delaney, Chuck, and Tom Schroeppel. *The Bare Bones Camera Course for Film and Video*, 3rd ed. New York: Allworth Press/ Skyhorse Publishing, 2015.

Dilley, Barbara. *This Very Moment: Teaching, Thinking, Dancing*. Boulder, CO: Dilley, 2015.

Doidge, Norman. *The Brain's Way of Healing: Remarkable Discoveries and Recoveries from the Frontiers of Neuroplasticity*. New York: Penguin, 2015.

Doidge, Norman. *The Brain That Changes Itself: Stories of Personal Triumph from the Frontiers of Brain Science*. New York: Penguin, 2007.

Ekman, Paul. *Emotions Revealed: Recognizing Faces and Feelings to Improve Communication and Emotional Life*, 2nd ed. New York: Henry Holt and Company, 2007.

Gershon, Michael. *The Second Brain: The Scientific Basis of Gut Instinct and a Groundbreaking New Understanding of Nervous Disorders of the Stomach and Intestines*. New York: Harper, 1998.

Hamilton, Diane Musho. *Everything Is Workable: A Zen Approach to Conflict Resolution*. Boston: Shambala Publications, 2008.

Holden, Lee. *Qigong for Health and Healing: A Complete Training Course to Unleash the Power of Your Life-Force Energy*. Boulder, CO: Sounds True, 2010.

Juhan, Deane. *Job's Body: A Handbook for Bodywork*, 3rd ed. Barrytown, NY: Station Hill Press, 2003.

Jung, Carl Gustaf. *Psychological Types. The Collected Works of C. G. Jung*, vol. 6, *Bollingen Series* XX. Princeton, NJ: Princeton University Press, 1976.

Kapit, Wynn, and Lawrence M. Elson. *The Anatomy Coloring Book*, 3rd ed. San Francisco, CA: Benjamin Cummings, 2002.

Kardam, Nükhet. *Ottoman to Turk and Beyond: Shimmering Threads of Identity*. Monterey, CA: Kardam, 2016.

Karr, Mary. *The Art of Memoir*. New York: Harper Perennial, 2016.

LaMothe, Kimerer. *A History of Theory and Method in the Study of Religion and Dance: Past, Present, and Future*. Leiden, The Netherlands: Brill, 2018.

Lang, James. *Small Teaching: Everyday Lessons from the Science of Learning*. San Francisco, CA: Jossey-Bass, 2016.

LeDoux, Joseph. *Anxious: Using the Brain to Understand and Treat Fear and Anxiety*. New York: Viking Books, 2015.

Levine, Peter. *In an Unspoken Voice: How the Body Releases Trauma and Restores Goodness*. Berkeley, CA: North Atlantic Books, 2010.

———. *Trauma and Memory: Brain and Body in a Search for the Living Past, A Practical Guide for Understanding and Working with Traumatic Memory*. Berkeley, CA: North Atlantic Books, 2015.

Linklater, Kristen. *Freeing the Natural Voice: Imagery and Art in the Practice of Voice and Language*. Hollywood, CA: Drama Publishers, 2006.

Lopez, Barry. *Horizon*. New York: Knopf, 2019.

Macrorie, Ken. *Telling Writing*, 4th ed. Boston, MA: Houghton Mifflin, 1985.

Maté, Gabor. *When the Body Says No: Exploring the Stress-Disease Connection*. Hoboken, New Jersey: Wiley; 2011.

Mayer, Emeron. *The Mind-Gut Connection: How the Hidden Conversation Within Our Bodies Impacts Our Mood, Our Choices, and Our Overall Health*. New York: Harper Wave, 2016.

McHose, Caryn, and Kevin Frank. *How Life Moves: Explorations in Meaning and Body Awareness*. Berkeley, CA: North Atlantic Press, 2006.

Menakem, Resmaa. *My Grandmother's Hands: Racialized Trauma and the Pathway to Mending Our Hearts and Bodies*. Las Vegas, NV: Central Recovery Press, 2017.

Myers, Tom. *Anatomy Trains: Myofascial Meridians for Manual Therapists and Movement Professionals*, 4th ed. Berkeley, CA: Elsevier, 2020.

Olsen, Andrea. *The Place of Dance: A Somatic Guide to Dancing and Dance Making*. Middletown, CT: Wesleyan University Press, 2014.

———. *Body and Earth: An Experiential Guide*. Middletown, CT: Wesleyan University Press, 2002/2019.

———. *BodyStories: A Guide to Experiential Anatomy*. Middletown, CT: Wesleyan University Press, 1991/2019.

Ornish, Dean, and Anne Ornish. *UnDo It! How Simple Lifestyle Changes Can Reverse Most Chronic Diseases*. New York: Ballantine Books, 2019.

Pert, Candace. *Molecules of Emotion: The Science Behind Mind-Body Medicine*. New York: Simon & Schuster, 1999.

Pollan, Michael. *In Defense of Food: An Eater's Manifesto*. New York: Penguin, 2009.

Porges, Stephen W. *The Polyvagal Theory: Neurophysiological Foundations of Emotions, Attachment, Communication, and Self-Regulation*. New York: W. W. Norton, 2011.

Quarry, Wendy, and Ricardo Ramirez. *Communication for Another Development*. London and New York: Zed Books, 2009.

Ray, Reginald. *Touching Enlightenment: Finding Realization in the Body*. Louisville, CO: Sounds True, 2014.

Rodrigues, Hillary, and John Harding. *Introduction to the Study of Religion*. New York: Routledge, 2009.

Saad, Layla F. *Me and White Supremacy: A 28-Day Challenge to Combat Racism, Change the World, and Become a Good Ancestor*. Naperville, IL: Sourcebooks, 2020.

Sadri, Houman A., and Madelyn Flammia. *Intercultural Communication: A New Approach to International Relations and Global Challenges*. New York: Continuum Books, 2011.

Scaer, Robert, MD. *8 Keys to Brain-Body Balance*. New York/London: W. W. Norton, 2012.

Siegel, Dan. *The Developing Mind: How Relationships and the Brain Interact to Shape Who We Are*, 3rd Edition. New York: Guilford Press, 2020.

Tomkins, Silvan. *Affect, Imagery, Consciousness: The Complete Edition*. New York: Springer Publishing, 2008.

Van der Kolk, Bessel. *The Body Keeps the Score: Brain, Mind, and Body in Healing of Trauma*. New York: Penguin Books, 2015.

VIDEO RESOURCES 👁

See www.wespress.org/reader-companions/.

Adler, Janet, Bonnie Bainbridge Cohen, and Nancy Stark Smith, "Why Here, Why Then?" and "Where Are We Now?" Two panels, produced by APE Gallery, Northampton Open Media, School for Contemporary Dance and Thought, and Historic Northampton, filmed at Smith College, Northampton, MA, Sept. 21, 2017, www. somatics2019.com/.

Hardwig, Scotty. "Andrea Olsen: Interview on Water." Produced by Middlebury College, Fund for Innovation. Interview and film by Scotty Hardwig, Apr. 5, 2018, video, 12:10, https://andrea-olsen.com/films/.

———. *Five Central Elements for Media Makers*. Produced by Middlebury College and the Middlebury Institute of International Studies and made possible with grants from the One Middlebury Fund and the Ron and Jessica Liebowitz Fund for Innovation, filmed in Carmel and Monterey, CA, 2018, 7:23, andrea-olsen.com/films/.

Hardwig, Scotty, Caryn McHose, and Andrea Olsen. *Body and Earth: Seven Web-Based Somatic Excursions*. Funded by the Digital Liberal Arts Initiative at Middlebury College. 2015, seven 10:00 videos. www.body-earth.org/.

Holden, Lee. "Qigong with Lee Holden." Produced by Middlebury College, Fund for Innovation. 2017, Santa Cruz, CA, video, 30:00. sites.middlebury.edu/wholebody/.

Kardam, Nükhet, "Watercolor Identities," filmed at TEDxMonterey, Monterey, CA, May 21, 2013, YouTube video, 18:01, www.youtube.com/watch?v=QvFOwb5ugWc.

Olsen, Andrea. "Moving from Fear to the Sublime: Art Making & the Environment," filmed at TEDxMonterey, Monterey, CA, May 14, 2011, 14:00, www.youtube.com/watch?v=Jmw2ZbLV-Hc.

Thorne, Gordon. "Vision of Thornes." Interview; film by Scotty Hardwig, produced by the Toad Foundation and the Northampton Historical Society, Northampton, MA, 2017, video, 13:00. www.apearts.org/gordon-thorne.html.

Subject Index

Page numbers in italics indicate images.

Abrams, Mary (Moving Body Resources), 134–38
addictive behavior/patterns, 169–70, 180, 189
Adler, Janet, xv, xviii, xxi, 31, 164, 200–4, *201*; Discipline of
    Authentic Movement, xv, 200–204
adrenal glands, 66–67, *66*
adrenaline, 66–67, 189
advertising, 93–94, 97–98
affects (inherited biological responses), 110, 134, 135, 135–36,
    137–38
Aiken, Chris, xii–xiii, *xiii*, 39, *128*, 130
alignment, xvii, 31, 36–43; agility, 36, 39–40, 112–13; in Body-Mind
    Centering, 194–95; cultural contexts, 40; with others, 40, 41–42,
    43, 131, 147, 200–201; postural, 36–39, 64, 77, 176; seated, 191,
    221. *See also* cultural heritage, values, and alignment,
Alvarez, Julia, 15, 87
ambivalence, 131, 132. *See also* lethargy and inertia
amygdalae, 56, *56*, 60, 62, 80–81, 168
animals: agility of, 36; communication and interaction with,
    13, 60, 116, 119, 120; as food, 178, 179; as part of landscape/
    locale, 32
animus, 169
architecture, 32, 104; sacred sites, 159–60, 163
arriving, 31–35, 145, 149
art-making, 99–106; communicating about, 104; creative process,
    100, 101–2, 105, 219–20; economics of, 104–5; intercultural
    collaboration, 100, 102, 103; natural/environmental, 101;
    social and environmental projects, 100, 103, 104
astrology, 15, 16
attention, 20; refreshing, 221
attitude, 5–8; positive, 7–8
aurora borealis (northern lights), 2, 101
Authentic Movement, xiv, 201, 202. *See also* Discipline of
    Authentic Movement
autonomic nervous system, xvii, 53, 54, 57–60, 66, 81, 136, 179;
    craniosacral division of, 59, 61
avatars, interactive, 92, 95
awareness, 143, 161, 185, 209; embodied/somatic, xv, xxi, xxii,
    25–26, 61, 137, 161, 194–95, 203; heightened, 161–63, 165; inner
    and outer, 200–201, 202; lens of perception, 9, 21, 25–27, 149,
    150. *See also* mindfulness practices; somatic (embodied)
    practices and therapies

babies (human infants): bonding and nurturing , 11, 48, 67, 76,
    110, 166, 173, 188; evolutionary heritage, 10, 11, 12, 54, 109–10,
    144, 195, 196; movement, 73; laughter and surprise, 122, 136;
    neurological development, 15; survival reflexes, 54, 55; voice
    and, 48

Eeva-Maria Mutka, Marloes Sands Beach, Wales (2015)
Body and Earth: Seven Web-Based Somatic Excursions
Still image from videography by Scotty Hardwig

balance, 36, 187; endocrine, 63–70; inner (centering), 189–90,
    194–99; nutritional, 63; vestibular system, 22. *See also* Body-
    Mind Centering
Bali, Indonesia, 161, 163; masks, *113*, *142*
beginnings, 31, 35, 78, 158, 164, 197, 202, 219. *See also* transitions
belly-mind (belly-brain), 46, 58, 97, 186–87, 189
betweenness, xv, xix, 3, 116
Blakeslee, Sandra and Matthew, xvii
blockages, unconscious: clearing, 73, 74, 129, 131, 133
body: attitudes and judgments about, 5–7, 74; biological
    inheritance (evolutionary development and programming),
    9–11, 12, 15, 49, 53, 54, 56, 195, 196
Body and Earth, 211–16
body image, 111, 179; non-verbal cues, 36, 109–11. *See also*
    self-image
body language (cues), 4, 7, 130–31, 145, 146, 152, 156. *See also* cues;
    stance
body map, 85, 92, 195
Body-Mind Centering, 194–99; School for, 194–95
body scan, 25–26, 189
bodystory, 15, 19
Bohm, David, 130
bones, 46, 80. *See also* alignment; posture; spine
boundaries, personal, 35, 129, 152, 156, 157, 170
bowing, 158, 161
brain: breathing and connectivity, 44, 46, 59; capacity, 11;
    cognitive processes, 56–57, 62, 68, 87, 88, 101–2, 144, 179,
    188–89; and computers, 92–93, 147; development, 11;
    neuroplasticity, 9, 136; relaxing, 167–68; response to humor
    and laughter, 122, 124; visual processing, 144–45, 146–47
brain, components of: amygdalae, 56, 60, 62; back (hind), 11,
    54, 188–89; cerebellum, *23*, 57, 62, 66, 88; cerebral cortex, 55,
    56–57, 101–2, 168, 188; corpus callosum, 56, *57*, 62; front, 11;

Caryn McHose, Fabiano Culora, Susanna Recchia, and Eeva-Maria Mutka, London, England (2015)
Body and Earth: Seven Web-Based Somatic Excursions
Still image from videography by Scotty Hardwig

Art Index

Monotype Parrots #4 (2018) by Philip Buller
Oil on resin paper, 44 in. × 51 in.

About the Author

ANDREA OLSEN, writer, performer, and interdisciplinary educator, is the author of three previous books on the body: *The Place of Dance: A Somatic Guide to Dancing and Dance Making* (with Caryn McHose; 2014), *Body and Earth: An Experiential Guide* (2002), and *BodyStories: A Guide to Experiential Anatomy* (1991). She is a professor emerita of Dance at Middlebury College in Vermont and was visiting faculty at the Middlebury Institute of International Studies at Monterey, California. She performs and teaches internationally and is a certified teacher of qigong. See andrea-olsen.com/.

Andrea Olsen
Photograph by Stephen Keith